NEW PERSPECTIVES IN FORAGE CROPS

Edited by **Ricardo Loiola Edvan**
and **Leilson Rocha Bezerra**

New Perspectives in Forage Crops
http://dx.doi.org/10.5772/66549
Edited by Ricardo Loiola Edvan and Leilson Rocha Bezerra

Contributors

Braulio Valles, Eliazar Ocaña-Zavaleta, Jesús Jarillo-Rodríguez, Epigmenio Castillo-Gallegos, Miguel Angel Alonso-Diaz, Edson Santos, Wilma Sá, Juliana Oliveira, Alexandre Perazzo, Ademar Pereira Serra, Mohammed Mouradi, Aurelio Pedroza, Luis Gerardo Yañez-Chavez, Adriana Cruz-Martínez, Ignacio Sánchez Cohen, Ricardo Trejo-Calzada, Jose Alfredo Samaniego-Gaxiola, Salah Attia-Ismail, Neil Anderson, Tereza Kávová, Barbora Kubátová, Vladislav Čurn, Allan Sebata, Atanas Sevov, Christina Georgieva Yancheva, Yanka Kazakova

Published by InTech
Janeza Trdine 9, 51000 Rijeka, Croatia

Publishing Process Manager Teo Kos
Technical Editor SPi Global
Cover InTech Design team

First published January, 2018
Printed in Croatia
Legal deposit, Croatia: National and University Library in Zagreb

Additional hard copies can be obtained from orders@intechopen.com

New Perspectives in Forage Crops, Edited by Ricardo Loiola Edvan and Leilson Rocha Bezerra
p. cm.
ISBN 978-953-51-3806-8

For EU product safety concerns:
IN TECH d.o.o., Prolaz Marije Krucifikse Kozulić 3, 51000 Rijeka, Croatia,
info@intechopen.com or visit our website at intechopen.com.

Meet the editors

Editor, Dr. Ricardo Loiola Edvan is a doctor in Animal Science from the Federal University of Ceará (2013). He has a master's degree in Animal Production from the Federal University of Paraíba (2010), with specialization in Agroecology from the Federal University of Paraíba (2008), and is a graduate in Agronomic Engineering from the Federal University of Ceará (2006). Currently, he is a professor of Animal Production, Agronomy, and Veterinary Medicine of the Federal University of Piauí acting in the area of forage and grassland and a professor of postgraduate programs in Animal Science. He is a professor who coordinates and participates in research and extension projects, articulated with public and private institutions.

Co-Editor, Dr. Leilson Rocha Bezerra has a PhD degree in Nutrition and Ruminant Production from the Federal University of Bahia (2015) and a doctor's degree in Animal Science from the Federal University of Paraíba (2009). He has a master's degree in Animal Nutrition from the Federal University of Campina Grande (2006), and is a graduate in Veterinary Medicine from the Federal University of Campina Grande (2005). Currently, he is a professor of Production and Nutrition of Ruminant Animals from Agronomy and Veterinary Medicine from the Federal University of Piauí. He is a professor of postgraduate programs in Animal Science at the Federal University of Piauí and Federal University of Campina Grande. He is a productivity researcher from the National Council for Scientific and Technological Development (CNPq-Brazil). He is a teacher who coordinates and participates in research and extension projects, articulated with public and private institutions.

Contents

Preface

Livestock farming is an activity of great importance to the world population, either through the production of food and products or through the generation of income. In some countries, livestock farming is responsible for the largest share of gross domestic product (GDP) produced in its territory and for a large part of people's food; in particular, livestock provides the main sources of protein for humans.

Livestock production systems depend on the production of food in quantity and quality for the animals. The most economically feasible form of food production for livestock is the supply to pasture, with grazing being the most recommended technique for most environments on our planet.

The production of fodder plants is undoubtedly the most efficient way to produce products of animal origin with quality and economic viability, due to the lower costs presented for this type of feed when compared to other feeding methods. Forage species and technologies used for forage production are very diverse and depend on the type of environment in which the animal production activity is to be carried out.

To produce forage with efficiency and quality, it is necessary to know the different variables that influence the growth of forage species and thus the production of feed for the animals. It is known that the elements of climate, soil, plant, and animal interact and affect the production of products of animal origin. The variables of climate and soil determine the growth, production, and quality of the forage that in turn influences the animal production in terms of quantity and quality, as well.

Understanding this complex and the interaction of these elements in the pasture environment is extremely important to maintain adequate production levels.

It is important to know the climate of the region in which the forage species is to be produced, especially in relation to rainfall and temperature, which directly influence the growth, production, and quality of the forage. In the soil, only fertility is of fundamental importance to maintain high plant production. The physical characteristics of the soil can also influence the growth of the forage species, and these factors also interfere in the quality of the fodder produced. To choose a suitable forage species, it is necessary to know the characteristics of the environment and the species available on the market. It is also necessary to consider the type of animal that will be used, so as not to choose a plant with characteristics inappropriate to the animal. Thus, disseminating information about forage species through the dissemination of studies and experiences of renowned researchers from different parts of the planet is certainly a relevant contribution to the area of animal production.

This book provides the reader with important information on developed researches and experiences of researchers with forage plants, offering relevant information on the subject of forage in different environments and about different forage species. The effect of the environment on the development of the forage species will be discussed in some chapters of this book. The use of fertilizers in the pasture environment is also the subject addressed in this book, as well as topics related to specific forage species.

The book "New Perspectives in Forage Crops" has 10 chapters by different experts in the area of forage farming, which deal with the themes related to the title of the book, especially on fertilization, forage production in the semiarid region, forage species selection, nitrogen fixation, grasses, legumes, cacti, drought, etc. The authors of the book are of different nationalities and thus provide a wealth of diverse information based on their diverse backgrounds. We hope readers will have a good reading and appreciate the information disclosed here by the authors.

Dr. Ricardo Loiola Edvan and Dr. Leilson Rocha Bezerra
Department of Animal Science,
Federal University of Piauí,
Brazil

Section 1

Forage Production in Dry Regions

Chapter 1

Water Harvesting and Soil Water Retention Practices for Forage Production in Degraded Areas in Arid Lands of Mexico

Aurelio Pedroza-Sandoval, Ricardo Trejo-Calzada,
Ignacio Sánchez-Cohen, Luis G. Yáñez-Chávez,
Adriana Cruz-Martínez and
Uriel Figueroa-Viramontes

Additional information is available at the end of the chapter

http://dx.doi.org/10.5772/intechopen.69618

Abstract

The area under arid conditions in Mexico is greater than 50%. This area faces a high risk due to environmental effects. The soil degradation in arid, semi-arid, and dry sub-humid areas is of multi-causal nature, among which climatic and anthropogenic factors stand out. At least, three distinct elements with different effects may be considered: recurrent droughts in short periods, long-term climate fluctuations, and degradation of soils by human activities. These threaten the productivity and sustainability of ecosystems and agro-ecosystems. Thus, it is needed to maintain a constant exploration of new and more appropriate technologies that promote the efficient use of natural resources, in a framework of greater sustainability. Many of these technologies are focused toward better management of water and soil resources in production systems. Water management is oriented with rainwater harvesting, efficient irrigation systems, as well as soil moisture retention techniques, and the use of plant species tolerant to water stress. Planting of native species and using soil improvers of edaphic moisture retention can enhance reclamation (recovery) of degraded soils. The aim of this chapter is to show and discuss some experimental results using the above technologies applied to rangelands with degraded soils in dry lands.

Keywords: soil degradation, dry lands, grasses, productive reconversion, soil water content

1. Introduction

Current world population exceeds 7200 million people and it is expected to reach 9.6 billion in 2050 [1]. The population growth is greatest in developing countries, with the consequent increase in pressure on the use of natural resources, in response to food demand and other services. Arid zones are among the top issues due to their territorial extension and the abundance and diversity of their natural resources. These regions represent one of the most viable options to respond to the challenges of the twenty-first century. However, there is a need for a systemic and comprehensive vision of the potential limitations that exist in these areas, which involves identification of current and future conditions on the state of natural resources, as a basis for the design and implementation of development plans. Globally, one of the challenges in this century is to reduce poverty in developing countries. One possibility to reach this goal is through a better management of natural resources involved in food production, such as water and soil, as well as the conservation of biodiversity, and the restoration of fragile ecosystems from the arid zones.

The socio-ecological systems in arid lands cover approximately 45% of the planet's surface, where about 2 billion people live, accounting for 33.8% of the world population [2–4]. In these regions, a critical situation is the depression of the local economies because of the impacts of the drought on economic activities and the natural resources, mainly water, soil, and native plants and animals, expressed in processes of desertification. These processes are expanding and Mexico will be one among the most affected countries. Thus, it is predicted that in 2050, rainfall and irrigated maize production will be reduced by 17 and 8%, respectively; while wheat production under rainfall systems could be reduced by 19% and in systems under irrigation by 8% [5].

The Mexican desert covers more than 50% of the national surface, which has different degrees of aridity typified by the characteristics of rainfall [6]. In this area, around 18% of the national population is located [7]. Moreover, the larger water demands are in the north and center of the country, which becomes a problem, since Mexico presents a territorial contrast of shortage and abundance of water [8]; while in southern Mexico frequent floods occur, in Northern Mexico, rainfall is very irregular and scarce. Then, the country is divided into two large areas: (1) Northern, central and northeast, where 77% of the population is concentrated, and 80% of gross domestic product (GDP) is generated in this region, but only 31% is renewable water. (2) Southern and southeast, where 23% of the population lives there, and only 20% of GDP is generated, but in this place, occurs 69% of renewable water [9].

2. Zoning for the best use of natural resources

A proper use of the natural resources in a region may be reached by a micro-regionalization, which allows one to know the amount of small areas that integrate any zone with natural similarities within each micro-region.

2.1. Agroecological zoning by geographical information system (AEZ/SIG)

The agroecological zoning is a methodology aimed at the evaluation of soil resources that integrates the use of geographic information systems. It has been applied in several countries and has been adopted as the method of evaluation of soil resources. This methodology is used in a holistic focus in territorial planning projects, and is a tool proposed by the AEZ methodology combined with the geographical information systems [10].

The development of cartographic procedures to obtain agroecological zoning is a current issue related to the concept of sustainability. These tools present the spatial classification of a territory per its potential capabilities, such as: production and protection zones, zone restoration, and identification of optimal areas for the intensive development of commercial plantations, among other uses. Thus, the decision takers may design plans to make better use and management of natural resources, such as biotic resource to increase profits, but also to improve and increase the existing natural capital over time.

AEZ/SIG methodology separates areas based on combinations of soil, physiography, and climate. This tool has several applications for land characteristics and types of uses such as potential land productivity, risk assessment of degradation, environmental impact assessment estimation of arable land, among others [11]. Most AEZ/SIG studies identify the types of land use (TLU) regarding crops or range of crops and input levels [12].

2.2. Case of study of zoning in arid lands

The middle watershed Nazas-Aguanaval region of the country is an area of the Chihuahuan desert, where ecosystems and agro-ecosystems with environmental, economical, and social importance are located. Beyond regionalization somewhat generic, the middle watershed Nazas-Aguanaval does not have a project of specific regionalization, based on an integrative approach from the physical and biological point of view, to allow better planning of natural resources directly related to regional development processes. This is important, given the need for greater agricultural production and, at the same time, to make more rational use of resources with minimal or no adverse environmental impacts. In this case of study [13], different indicators were used to make the comprehensive characterization of a region, considering both physical factors and the most important biological factor [14]. Aridity index, index of seasonal drought, vegetation, and type of soil were the indicators used for the survey of regionalization. Steps of study: (1) The middle watershed Nazas-Aguanaval map was generated with digital sub-watersheds [15] and the boundaries that are defined between the Francisco Zarco and Lázaro Cárdenas dams, shunt, and storage of RH 36, respectively (**Figure 1A**); (2) From the National Weather Service, we obtained the climatic stations for the middle watershed Nazas-Aguanaval of the states of Durango and Zacatecas, Mexico, which were geo-referenced (**Figure 1B**); (3) Applying the program [16] and the INEGI data base [15], the physiographic (vegetation and soil) characteristics of the middle watershed Nazas-Aguanaval were identified, using the digital information from the native vegetation and soil scale 1:250,000 (**Figure 2A** and **B**); (4) Interpolation was performed by the inverse distance weighted (IDW) method with the ArcMapTM 10.1 software to obtain the drought and aridity index using a raster graphics.

The different micro-regions were identified based on the degree of aridity. We used the Emberger aridity index modified by Stretta and Mosiño [17], which integrates the predominant action of the rainfall regime and the influence of the maximum and minimum average temperatures of the hottest and coldest, respectively. For the calculation of the aridity condition, the series of historical climatological data (1979–2008 of the National Meteorological Service) of 26 climatic stations in the study area was used (**Figure 3A**).

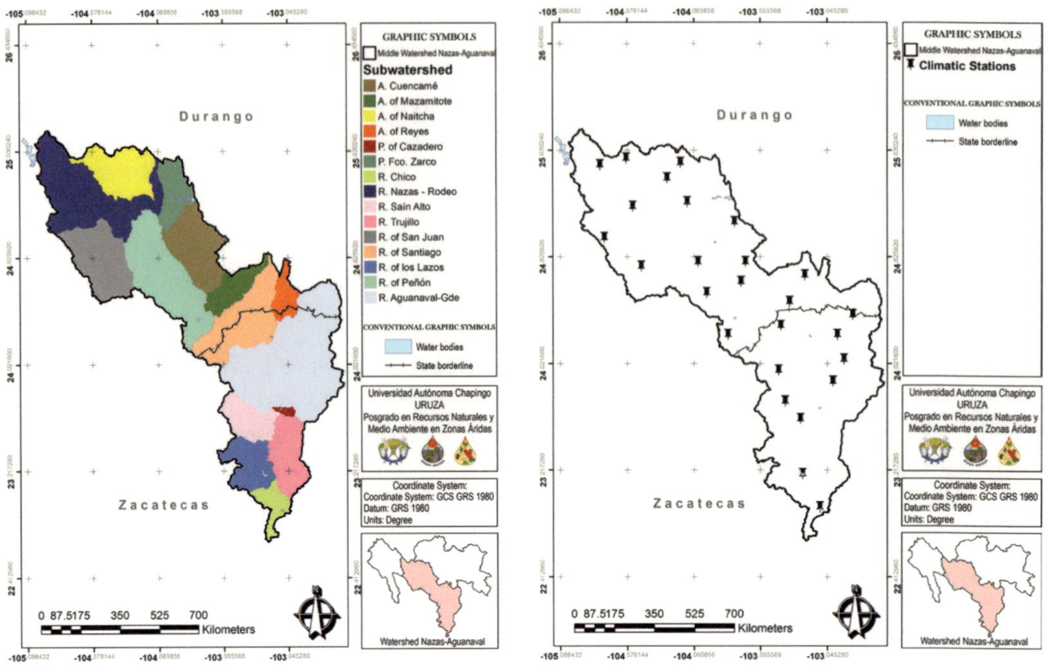

Figure 1. Middle watershed Nazas-Aguanaval and sub-watersheds (A) geo-referencing climatic stations, marked with black dots (B) (Source: Compiled by author based on maps available at [15]).

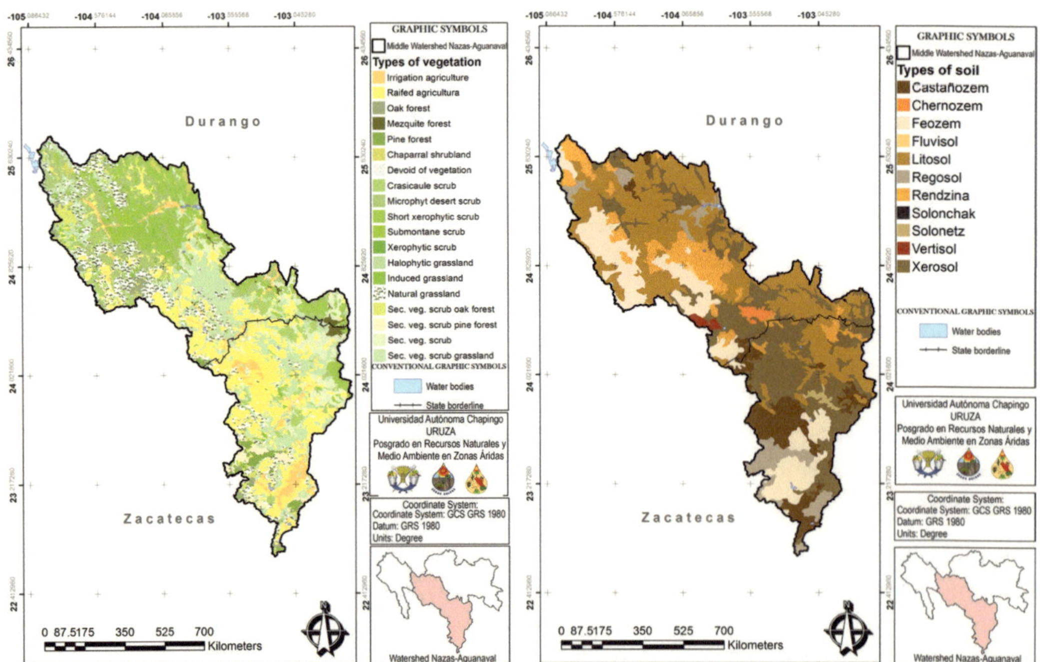

Figure 2. Types of vegetation (A) and types of soil (B) in the middle watershed Nazas-Aguanaval (Source: Compiled by author based on maps available at [15]).

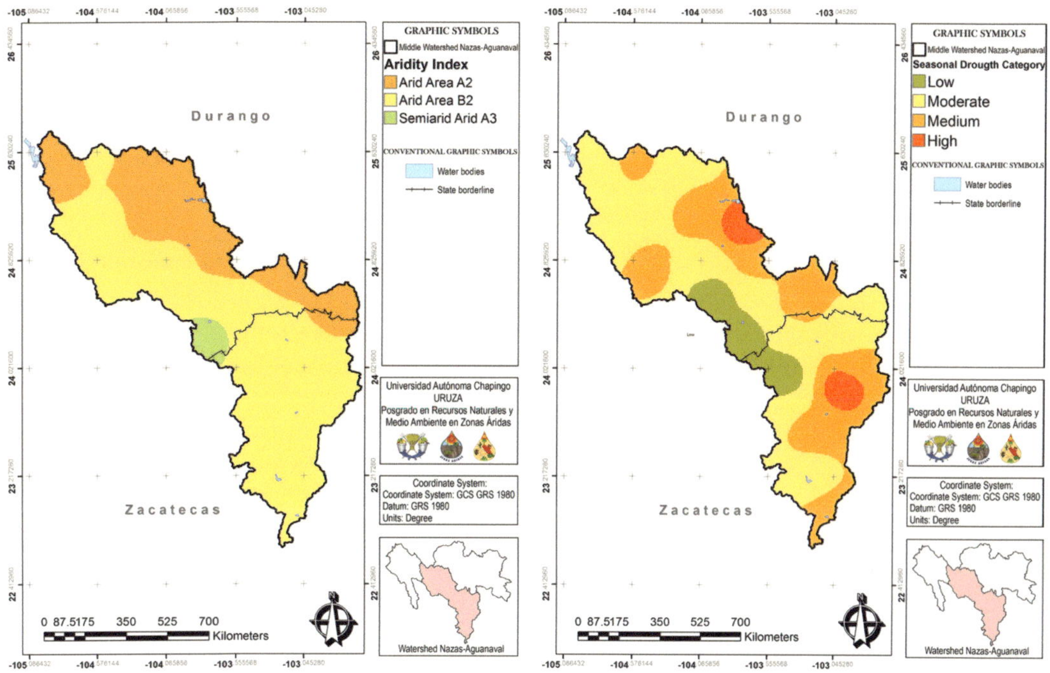

Figure 3. Levels of aridity (A) and spatial distribution of seasonal drought in the middle watershed Nazas-Aguanaval (B). (Source: Compiled by authors).

Annual drought regionalization and its intensity were calculated through the water precipitation information, which was collected from the 26 climactic stations with complete data for every year [18]. The number of years with drought was recorded for the selected seasons. A scale in percentage of years in the presence of drought was established to categorize the levels (**Figure 3B**).

The categorization proposed by integrating the indicators mentioned above arises from the possible combinations of the different levels of each regionalization approach, resulting in the following micro-regions:

Micro-region 1: corresponding to the most arid place of the aridity index type A_2, with predominant halophytic grassland vegetation and moderate seasonal drought. **Micro-region 2**: is the most arid place of the aridity index type B_2, with natural grassland having a seasonal medium drought. **Micro-region 3:** corresponding to less arid of the aridity index A_2, with microphyll desert scrubland but with a moderate seasonal drought. **Micro-region 4:** also corresponding to less aridity index A_2, but with predominant halophytic grassland vegetation and rainfall agriculture, and moderate seasonal drought. **Micro-region 5: also corresponds** to most aridity index A_2, but with xeric scrubland as main vegetation and short xerophytic scrub as secondary vegetation, but with a medium seasonal drought. **Micro-region 6**: corresponding to aridity index A_2, like the previous micro-regions, but here the predominant native vegetation is microphyll desert scrubland and xeric scrubland as secondary vegetation, and has a high seasonal drought. **Micro-region 7**: with aridity index B_2, with natural grassland as vegetation primary and chaparral scrubland as secondary vegetation and medium seasonal drought. **Micro-region 8**: also with aridity index B_2, but with xeric scrubland as main vegetation and natural grassland as secondary vegetation, but with a moderate seasonal drought. **Micro-region 9**: with the same aridity index B_2, but here the natural grassland as the main vegetation and rainfall agriculture and a low seasonal drought. **Micro-region 10**: with aridity index A_2, but with halophytic grassland as main vegetation and xeric scrubland as secondary vegetation, but with a medium seasonal drought. **Micro-region 11**: also aridity index B_2, but predominantly crassicaule scrubland vegetation and a medium seasonal drought. **Micro-region 12**: with aridity index A_3, corresponding to less arid place from the semiarid areas, with rainfall agriculture as main vegetation and pine forest as secondary vegetation, but with a low seasonal drought. **Micro-region 13**: aridity index A_2 with xeric scrubland having a moderate seasonal drought. **Micro-region 14**: corresponding to aridity index B_2, but predominantly natural grassland vegetation and a

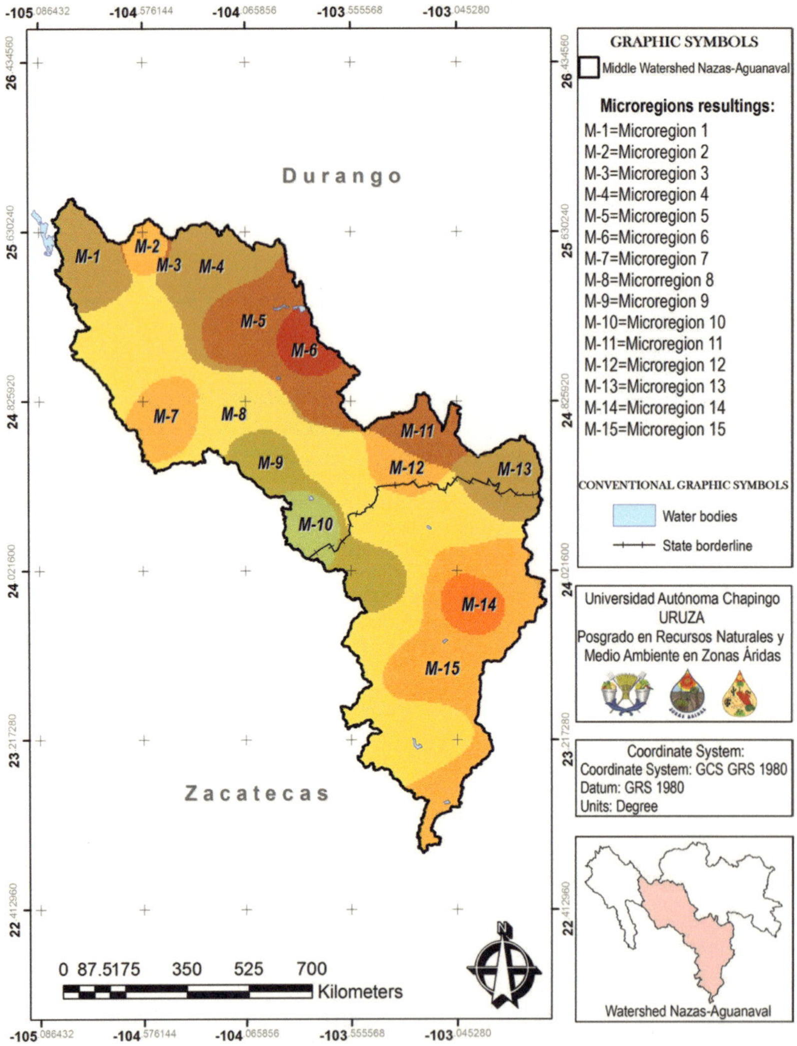

Figure 4. Micro-regions resulting in the central-north region of Durango by type of vegetation and soil, degree of aridity and intensity of temporary drought (Source: Compiled by authors).

high seasonal drought. **Micro-region 15**: aridity index B_2 with rainfall agriculture having a medium seasonal drought (**Figure 3**).

The identification of the 15 micro-regions of the study area, should allow planning a more targeted and systematic use of natural resources, based on the biophysical characteristics. A planning process with different purposes, such as: conservation, development and/or production of biotic resources in arid zones. According to different authors [19–21], regionalization is the best planning tool, especially when integrating different biophysical indicators such as the type of soil, type of native vegetation, frequency of droughts and aridity conditions (**Figure 4**).

3. Water harvesting and soil water conservation practices in arid lands for forage production

Through soils with low productive capacity and crops with high resistance to water stress could be possible to define areas for productive reconversion. The criteria include water harvesting and the use of some humidity retainers to increase the water holding capacity of the soil. Of this way is possible to promote the gradual recovery of the soil cover reducing soil and wind erosion, which are the main problems related to soil productivity in arid lands.

3.1. Degradation of soil in arid zones as the problem to mitigate

Desertification occurs because of the degradation of natural ecosystems in dry lands, which is a big global problem. Soil degradation causes may be included into three broad aspects: (a) physical, where climate plays a big role in terms of floods, droughts that enhance soil erosion; (b) chemical, generally in the form of salinization; and (c) biological, mainly as a result of the oxidation of the organic matter of the soil [22].

The main consequences of land degradation are chemical degradation of the soil, related to soil chemical contamination with several toxic elements like heavy metals and salts, which makes loss of vegetative cover, loss of soil surface layer infiltration, reduction of water storage in the soil, loss of soil organic matter, fertility and structure, loss of soil elasticity, loss of natural regeneration, and decrease of water level. Degradation affects most of the arid zones, mainly in marginal cultivated areas [11].

Most of the world's dry lands are degraded. In Mexico, estimates of the magnitude of degradation may differ from the methods used to calculate them. Even though, there are not specific studies on the extent of desertification in Mexico, the approach of this chapter considers soil degradation as an estimator of desertification recognizing the limitation, since only one considers its elements. On the other hand, the information included data back to 10 years [23].

3.2. Decision-making natural resource management

Before making any practices for planting native vegetation species that rehabilitate degraded soils in arid zones, are necessary to select wisely the species to grow. The aim of this part of study was to select species that meet the environmental conditions. Once the main problem is identified, the resource management system (RMS) planning tool provides a series of alternatives to cope with those problems. These alternatives were adjusted according the specific conditions encountered in the field [24].

The usefulness of the RMS planning tool is to perform a decision matrix composed by the resource concern and the alternatives. The soil, water, animals, plants atmosphere humans (SWAPAH) serves this purpose after a series of problems have been identified in the field. The tool provides a series of alternatives that are evaluated according to its impact on the resource concern. Therefore, a group of professionals select those alternatives with the less negative impact (or the more positive impact) on the resource. These selected alternatives are then emigrated to a decision support system, which are evaluated according to several criteria imposed by the participants (professionals and those users directly impacted by the decision). For standardizing the scores from 0 to 1 given to the criteria, a scoring function should be selected among: more is better linear, more is better nonlinear, more is worst linear, and desirable range [25, 26]. To facilitate the use of the software, the user is presented with the shapes of these functions from where one should be selected [25, 26].

Physical, social, economic, and environmental factors were included in this methodology. The best evaluated practices for their feasibility, viability, and ease of implementation were: grow woody plants (*Prosopis* spp) and grazing species, particularly buffel grass (*Cenchrus ciliaris* L.), which is able to adapt to the arid conditions [24].

4. Recover degraded soil in arid lands: some technologies

4.1. Practices for rain water harvesting and soil moisture retention to plant grass

Drylands are of high ecological vulnerability due to low vegetative cover and erratic and torrential rainfall. The aim of this study was to evaluate the use of different sources and dosages of soil moisture retainers in planting buffel grass (*Cenchrus ciliaris* L) in a micro-watershed system for harvesting rainwater. Micro-watershed system, are spatial units of different dimensions, since the smallest like 1 m^2 or larger dimensions, depending on the specific conditions of the site and based mainly on the hydrological concept of soil division (**Figure 5**). Additionally, in this survey were used two types of soil water retainers: one chemical product

Figure 5. Micro-watershed for water harvesting; when was built (A), and after, when the grass has grown (B).

named hydrogel, and the other was organic compost. Hydrogel is a polyacrylamide copolymer that, when used in the soil as a substrate, absorbs and retains large amounts of moisture and nutrients, making them available to the plant. In this survey, four dosages of hydrogel were evaluated: 0, 5, 10, and 15 kg ha^{-1} as well as two compost dosages: 0 and 40 t ha^{-1} based on dry weight, in the catch system of rainfall water as an experimental unit of 2 m^2 each one. The hydrogel and organic compost were incorporated to the soil manually [27].

4.2. Moisture content in soil

At 241 days after planting (DAP), the soil moisture content was higher, and then decreased to 346 and 372 DAP, with average values of 22.5, 16.8, and 8.2%, respectively (**Figure 6**). This trend is related to the rainfall with drought during May and June of the study year, followed by a period, also with relative drought during July and August, but in general this year (2013) was lack of rainfall since the annual average was only 205.4 mm (**Figure 7**) [28]. Although the response function showed a negative rate, and showed differences at all three sampling dates, the soil moisture content was always higher when the hydrogel was applied at least for the first two sampling dates (241 and 346 DAP), with values on the first date (241 DAP) of 25, 23.2, and 23.4% when applying 5, 10, and 15 kg ha^{-1}, respectively, vs 17.5 % when the hydrogel was not applied. A similar pattern was shown in the second sampling date (346 DAP); while at 372 DAP, which was not affected by moisture retention, and the different dosages of hydrogel (**Figure 8**). This means that about 157 days after the first sample without a significant rainfall, the effect of soil moisture retention of the hydrogel stopped, arising near the soil wilting point (WP) level. Thus the effect of the hydrogel was identified at least in the first two samples in any dosages included in this study, and the lower dosages of hydrogel (5–10 kg ha^{-1}) may be used to obtain the same results. Before partially is disagree to [29], who used dosages of 0, 20, 30 and 40 g of hydrogel in 130 g of sand, the higher dosages of hydrogel showed the best results as retainers of soil moisture [24].

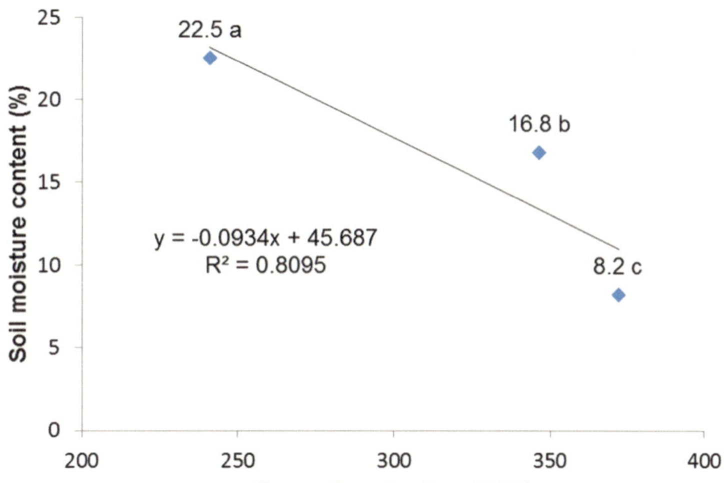

Figure 6. Soil moisture content at different sampling dates in days after planting (DAP) (2013).

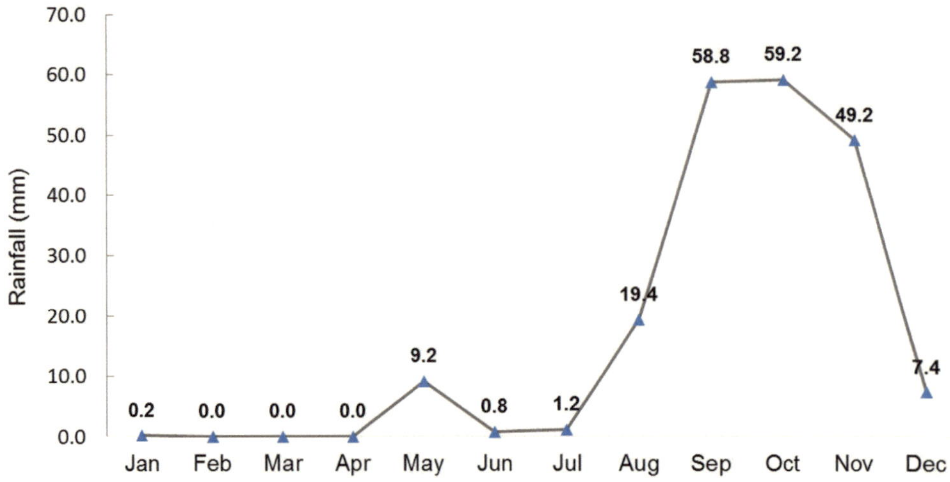

Figure 7. Rainfall during 2013 in the area near to the experimental area to Bermejillo, Dgo, México.

Seedling emergence was 47.7%, when 15 kg ha^{-1} of hydrogel was applied, compared to 29% when it was not applied. The dosages of 5 and 10 kg ha^{-1} had intermediate values, with no statistical difference (**Figure 9**). These results are consistent with those reported by Rojas et al. [30] regarding the use of hydrogel, which had a positive effect on the capacity of germination of tomato (*Lycopersicon esculentum* Mill).

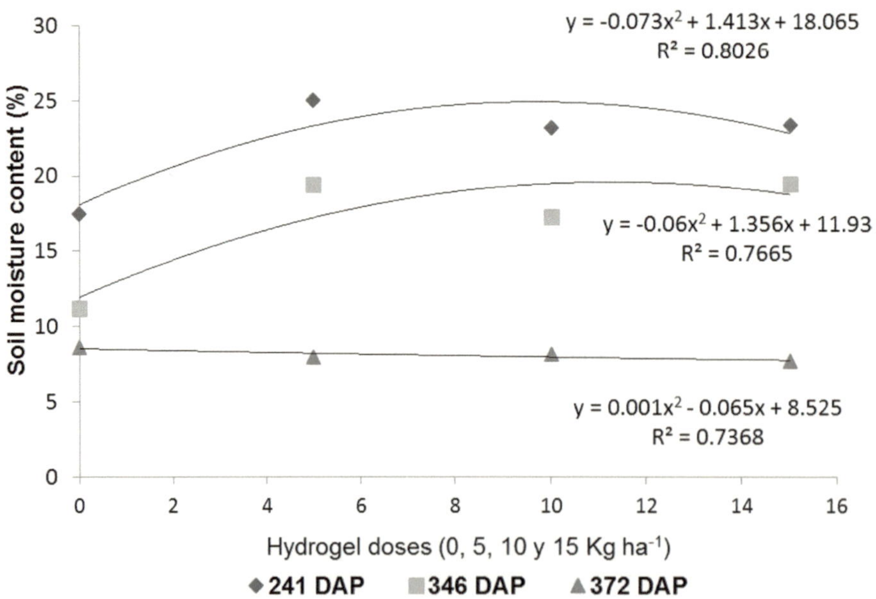

Figure 8. Soil moisture content with different dosages of hydrogel and sampling dates (DAP) (2013).

http://dx.doi.org/10.5772/intechopen.69618

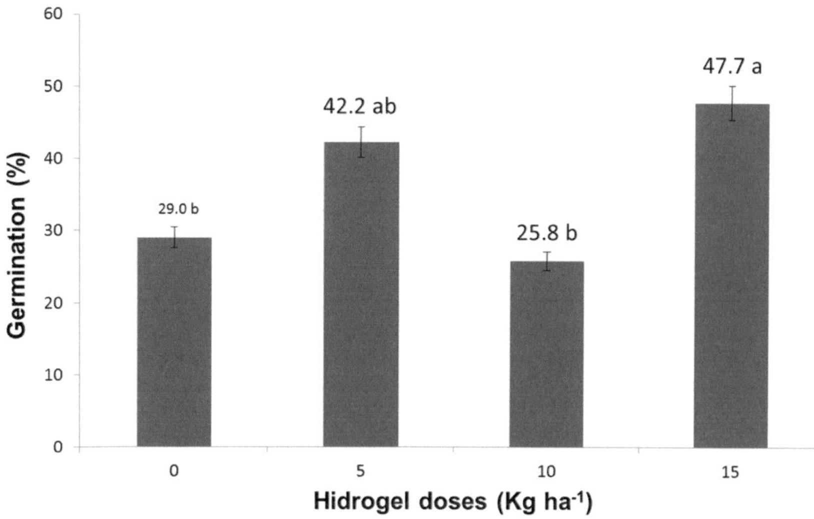

Figure 9. Effect of hydrogel dosages on germination percentage of buffel grass in micro-watershed for water harvesting, during August, 2012.

The lower dosages of hydrogel (5–10 kg ha^{-1}) used in this study were less effective in seedling emergence, which could mean that at this stage of development of buffel grass, high moisture content is required in the soil, which is expressed to a dose of 15 kg ha^{-1}. There was no effect on the compost on seed germination of grasses. Even with the low rainfall regime in 2012 and the absence of the treatment effect on the moisture content in the soil, the growth of buffel grass remained constant, with significant growth in each evaluation date, at an exponential rate of 1.7 cm d^{-1} (**Figure 10A**). This is an indicator of the high adaptability of this forage species under drought [31, 32]. Similar behavior was observed with grass development during 2013, with an exponential growth rate of 1.5 cm d^{-1} (**Figure 10B**). The height of the plant was significantly major in both dosages of hydrogel and consequently the dry matter weight too.

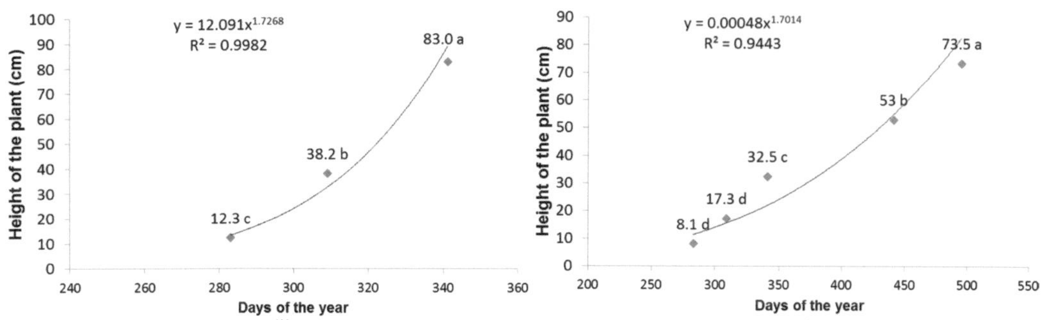

Figure 10. Growth of buffel grass in rainwater capturing watersheds, at different evaluation times during 2012 (A) and 2013 (B).

Hydrogel doses	Soil moisture content (%)	Photosynthesis (μmol m^{-2} s^{-1})	Conductance (mol m^{-2} s^{-1})	Transpiration (mmol H_2O m^{-2} s^{-1})
0	16.4b	3.72b	0.0055b	0.214c
5	19.5a	3.82b	0.0089a	0.227bc
10	17.2ab	4.05ab	0.0066ab	0.255b
15	18.4a	6.67a	0.0099a	0.382a

Tukey test ($P \leq 0.05$). Figures with same letters within the same column are not statistically different.

Table 1. Effect of hydrogel on photosynthetic activity and other physiological variables on buffel grass (*Cenchrus ciliiaris* L.).

Development buffel grass is dependent on the soil water content, but is tolerant to drought stress [33]. In this sense, hydrogel showed better effects enhancing plant growth and increasing biomass production in other crops as beet (*Beta vulgaris* var. cycla) [34] and tomato (*Lycopersicon esculentum* Mill) [35].

4.3. Photosynthetic activity and transpiration

Photosynthetic activity and transpiration of the plant were measured by the $C0_2$and H_20 flow, respectively using the infrared gas analysis (IRGA) model LICOR-6400. The photosynthesis was significantly higher in the dosages of 15 kg ha^{-1} of hydrogel at a rate of assimilation 6.67 mmol CO_2 m^{-2} s^{-1}, compared to the assimilation rate obtained when the product was applied to 5 or 10 kg ha^{-1} or not applied, with values of 3.7, 3.8, and 4, respectively. Higher photosynthesis rates were associated with greater conductance and transpiration, and vice versa (**Table 1**). Thus, the presence of moisture in the soil promotes plant photosynthetic activity, while water deficits decrease it [36]. The photosynthetic activity of the grass was strongly influenced by the condition of soil moisture; this was identified on November 6, 2013, with an average soil moisture of 18.4% when the hydrogel was applied, with no statistical differences between the dosages. All this indicates that the moisture content in the soil is influenced by hydrogel (**Table 1**).

4.4. Another practice of retention of soil moisture

The aim of this study was to evaluate different soil moisture retention practices in the survival and growing grass (*Boutelova curtipendula* and *Chloris gayana*) in areas of productive reconversion. Dosages of stubble of dry corn (0 and 10 t ha^{-1}), and hydrogel dosages (0, 10 and 20 kg ha^{-1}) were used, including two species of grasses *B. curtipendula* (native grass) and *C. gayana* (introduced grass) [37].

4.5. Seed germination

A number of seeds germinated *B. curtipendula* was to a rate significantly higher than the introduced species *C. gayana*. Germination started 2 days after sowing at a rate of germination in a logarithmic function (**Figure 11**). The latency mechanism of the native species is inherent to its

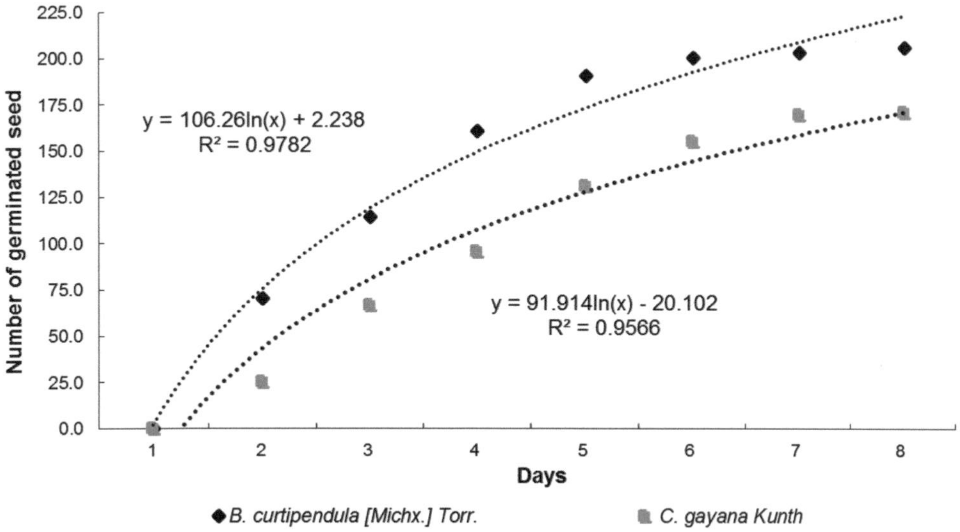

Figure 11. Germination rate in pastures *B. curtipendula* [Michx.] Torr. and *C. gayana* Kunth; discontinuous points represent the trend of the line.

high germination capacity [38]; even a high percentage of germination of exotic species is not enough for the success in potential production. Additionally, one characteristic of grass *B. curtipendula* was the high germination rate in the early stages of growth [39].

4.6. Moisture content

The moisture content was higher when applying hydrogel at 15 days after the rain (DAR) at each soil depth evaluated. This effect disappeared after evaluation dates without significance among hydrogel treatments for both soil depths (**Figure 12**). Hydrogel offers better water

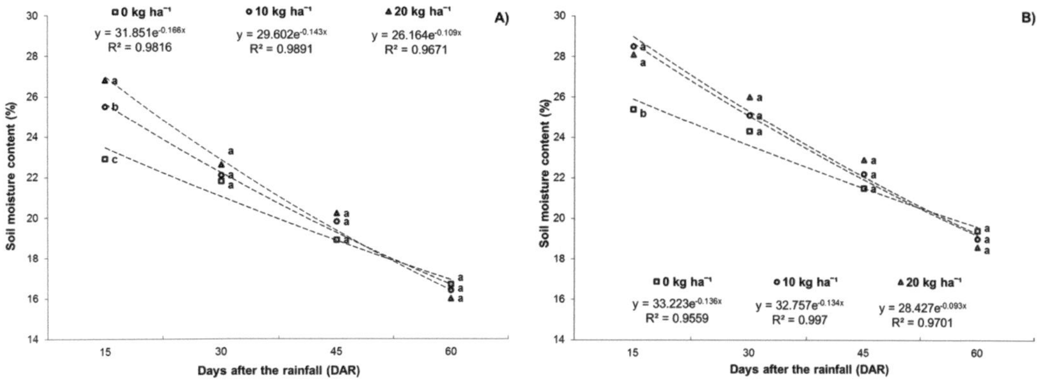

Figure 12. Moisture reduction at 30 cm (A) and 45 cm (B) depth of soil in different hydrogel contents and different sampling dates: 15, 30, 45, and 60 days after the rainfall. Tukey test ($P \leq 0.05$). Different letters over the same line means differences among treatments.

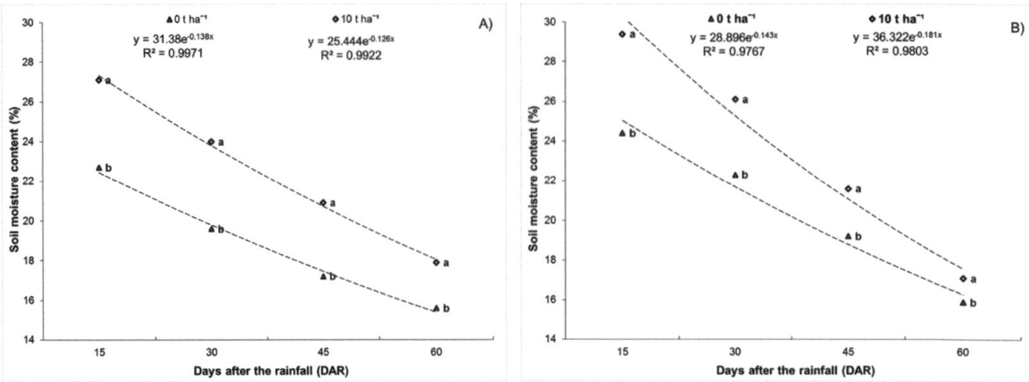

Figure 13. Moisture reduction at 30 cm (A) and 45 cm (B) depth of soil associated with dry corn stubble doses and different sampling dates: 15, 30, and 45 and 60 days after rainfall. Tukey test ($P \leq 0.05$). Different letters over the same line means differences among treatments.

release properties when combined with soil and retains larger amounts of water, either under instant or prolonged irrigation conditions, and maintains moisture at a higher value in crops [40]. These results differ from those found for other research studies [41], who found higher soil moisture content when applying hydrogel to a soil depth of 0 – 15 cm, in relation to the control, throughout the rice growing season. Also the hydrogel improves the rapidly available water capacity (RAWC) of soils; although the effectiveness of the gel in improving soil water retention varies for different soil types [42].

Moisture retention characteristics of the hydrogel are inherent to constant hydration, a condition that was not present at the study site due to low rainfall; in other hand, the occurrence of precipitation for arid and semi-arid zones is not homogeneous, and highly variable in time. Regarding the application of dry corn stubble, the moisture content at both evaluated depths remained higher at each evaluation date than the harvest residue was not added (**Figure 13**). The useful moisture range for this type of soil is 18%, since the field capacity (FC) is 33% and the permanent wilting point (PWP) is 15%. In the treatment without stubble application reached values of 16% always, very close to PWP. Application of dry corn stubble maintained soil moisture content higher than 18% always, with a lower rate of humidification than the treatment without stubble an average of both soil depths (30 and 40 cm); the treatment with corn stubble obtained 3.7 and 3.1% more moisture content than the control at 30 cm and 45 cm depth, respectively.

4.7. Percentage of survival

The percentage of survival of the grass was significantly different from the species (**Figure 14**). Both *B. curtipendula* and *C. gayana* had a survival rate greater than 84%, 6 weeks after transplant; however, *C. gayana* had a higher percentage with 87% compared to the native grass, which reported 84.1%. Agree Ref. [43], the grazing transplant method represents an effective technique to increase the percentage of survival in grassland areas. The practice of planting

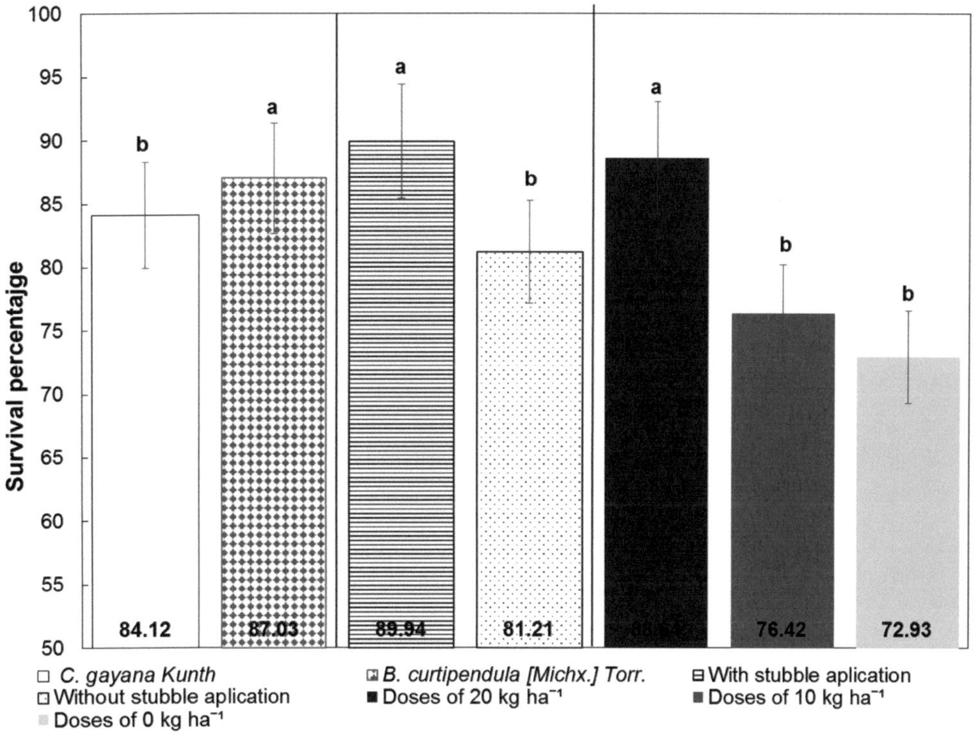

Figure 14. Percentage of survival by grass species with application of 10 t $ha^{-1}$of stubble and without application of stubble; application of hydrogel at doses of 20 kg ha^{-1}, 10 kg ha^{-1}, and 0 kg ha^{-1}. Vertical bars represent the standard deviation ±. Columns with equal letters are not statistically different ($P \leq 0.05$).

seeds of forage grasses improves the productivity of livestock areas in the arid and semi-arid zones. Studies have shown that the practice of planting directly has success of only 10%; in other cases, the probability of success is 50%. In contrast, for this study, values higher than 95% of live plants ([number of deaths plants/number of live plants] × 100, after 45 days of the transplanted) in buffel grass was shown using the transplant method, considered a highly effective method of planting even in soils with limited natural fertility [43]. In addition, it has been reported that, for experimental procedures, the transplantation technique guarantees the obtaining of reliable information by reducing the experimental error to acceptable values. The percentage of planting buffel grass varied from 74 to 99%, without significant differences between plant species, which had an average of 90% of establishment [44].

The percentage of survival of the grasses when applying hydrogel was significantly higher, compared to no application; 89.3% and 76.4% of survival were found for the doses 20 and 10 kg ha^{-1}, respectively, without statistical difference between them. Similar results showed treatments with stubble application, the percentage of survival was significantly higher, when applying corn stubble (89.9%), with respect to that when it was not applied (81.2%) (**Figure 13**). Similar results have been reported, when vegetation cover is not applied, the percentage of forest species survival is significantly reduced to 66.7% [45].

Dose of retainer of soil water	Number of tillers		Plant height (cm)		CCI		Vigor (0–5)	
	BC	CG	BC	CG	BC	CG	BC	CG
Hydrogel								
0 kg ha^{-1}	5.6a ±1.0	3.5a ±0.8	26.9a ±4.8	33.1a ±4.8	98.6a ±5.3	114.3a ±5.1	3.7a ±0.5	4.1a ±0.6
10 kg ha^{-1}	5.7a ±0.9	3.6a ±0.9	27.7a ±5.2	34.9a ±4.6	95.9a ±4.7	113.2a ±5.7	3.3a ±0.7	4.5a ±0.7
20 kg ha^{-1}	5.9a ±1.1	3.2a ±0.9	27.2a ±4.3	32.6a ±3.6	97.8a ±4.3	109.8a ±6.2	3.5 a ±0.4	4.3a ±0.5
Dry stubble coverage								
0 t ha^{-1}	4.4b ±0.9	3.0b ±0.7	22.6b ±2.6	29.5b ±3.1	84.4b ±6.4	107.8b ±6.7	4.0± 0.4 [b]	3.8b ±0.5
10 t ha^{-1}	7.1a ±0.8	3.8a ±0.8	31.9 a ±3.3	37.2a ±3.4	98.5a ±5.9	121.5a ±7.4	2.9a ±0.7	4.7a ±0.3

BC = *B. curtipendula* [Michx.] Torr.; CG = *Chloris gayana* Kunth; CCI = Chlorophyll Content Index. [ab]Numbers of different letter into the same column, and into each variation factor (hydrogel and dry stubble coverage) are statistical different ($P < 0.05$).

Table 2. Growing variables in two grass species in different hydrogel doses and dry stubble coverage.

4.8. Growth and plant development

Dry stubble coverage at doses 10 t ha^{-1} significantly influenced the plant height, number of tillers, and chlorophyll index and plant vigor of the grass species during different evaluation dates (**Table 2**). The introduced grass was superior in all variables with respect to the native grass. Studies with stubble use showed effects with the addition of mulch or peat cover on peanut cultivation, significantly influencing some agronomic attributes of crop growth and yield [46].

Plant growth, forage yield of the guinea grass was increased when straw in saline soils was applied [47]. Also Ref. [48] identified that the incorporation of mulch to the soil for establishment of grasses, had superior results, related to maximized vegetation of grasses and greater amount of biomass, regarding the treatments without addition of straw.

4.9. Biomass production

The effect of stubble was related to higher and evenly distributed soil moisture content, which allowed a better yield of biomass in each evaluation. In contrast, the hydrogel showed only statistical differences at 30 days after the transplantation when applying 10 or 20 ton ha^{-1}, without statistical difference in the first dose with the control, in the case of *B. curtipendula*. In the two subsequent evaluations, the effect was no longer shown, which may be related to the dilution effect identified in soil moisture content. The introduced grass was always higher in area and radicular biomass for the three evaluations performed (**Table 3**).

Dose of retainer of soil water	Dry weight of biomass (g)					
	30 DAT		45 DAT		60 DAT	
	BC	CG	BC	CG	BC	CG
Hydrogel						
0 kg ha^{-1}	11.0b ±1.7	17.5b ±3.3	45.1a ±6.0	57.9a ±5.5	53.3a ±7.1	68.0a ±7.0
10 kg ha^{-1}	13.3ab ±2.2	24.3a ±4.1	42.8a ±5.9	56.0a ±5.8	51.5a ±6.3	66.8a ±7.6
20 kg ha^{-1}	14.7a ±2.0	25.6a ±4.9	44.7a ±5.3	56.9a ±4.2	51.6a ±6.2	67.7a ±5.8
Dry stubble corn						
0 t ha^{-1}	12.3b ±2.1	20.4b ±1.6	39.8b ±3.3	52.3b ±2.7	47.2b ±4.4	61.1b ±2.8
10 t ha^{-1}	16.1a ±0.9	28.6a ±1.9	48.8a ±3.9	61.1a ±2.8	57.2a ±4.5	73.2a ±3.2

BC = *B. curtipendula* [Michx.] Torr.; CG = *Chloris gayana* Kunth DAT = Days after of the transplant. [ab]Numbers of different letter into the same column, and into each variation factor (hydrogel and dry stubble coverage) are statistical different ($P < 0.05$).

Table 3. Biomass production in two grass species to different hydrogel and dry stubble corn dosages.

5. General conclusions

Ecological zoning through indicators such as aridity, drought, type of vegetation, and type of soil use is often an effective tool in a more systematic and targeted application of technologies for a better management of natural resources, according to the potential of each small region.

Water is the most limited natural resource and therefore of major importance to be considered in development plans in arid lands to optimize the use of this resource and avoid environmental deterioration such as soil degradation, turning productive areas into unproductive ones.

An integrated system, such as use of water stress tolerant plant species, rainwater harvesting, and soil moisture retention practices, like the experiences and results shown in this chapter, may be the useful tools for the productive reconversion of areas degraded in arid lands.

Exploration and integrating technological practices focused on promoting effective management of natural resources in ecosystems and agro-ecosystems, is a current an issue where the main constraint is water.

Acknowledgements

We want to thank to Chapingo Autonomous University for financial support to carry out the research related to the topic of the mitigation of degraded soils in arid zones, as well as to the National Institute of Agricultural Livestock and Forestry Research for their collaborations in this project.

Author details

Aurelio Pedroza-Sandoval[1], Ricardo Trejo-Calzada[1*], Ignacio Sánchez-Cohen[2], Luis G. Yáñez-Chávez[1], Adriana Cruz-Martínez[1] and Uriel Figueroa-Viramontes[2]

*Address all correspondence to: rtrejo@chapingo.uruza.edu.mx

1 Chapingo Autonomous University, Bermejillo, Durango, Mexico

2 National Institute of Agricultural Livestock and Forestry Research, Ciudad de México, Mexico

References

[1] ONU. La población mundial alcanzará los 9.6 mil millones en 40 años [Internet]. 2014. Available from: http://actualidad.rt.com/actualidad/view/97314-poblacion-mundial-crecimiento-onu [Accessed: 6 March 2016]

[2] Cruse MR. Agriculture: Is climate change a serious issue? Revista Agrociencia Uruguay. Special Issue. 2012; Sept.:9-19

[3] Al-Kaisi M, Elmore RW, Guzman JG, Hanna HM, Hart CE, Helmers MJ, Hodgson EW, Lenssen AW, Mallarino AP, Robertson AE, Sawyer JE. Drought impact on crop production and the soil environment: 2012 experiences from Iowa. Journal of Soil and Water Conservation. 2012;**68**(1):19-24

[4] UNCCD. Segunda Conferencia Científica de la CLD; Aspectos económicos de la desertificación, la degradación de las tierras y la sequía: Metodologías y análisis para la toma de decisiones [Internet]. 2013. Available from: http://2sc.unccd.int/es/inicio [Accessed: 14 March 2014]

[5] IFPRI. Informe de Políticas Alimentarias Mundiales 2011. Washington, D.C.: IFPRI; 2012. p. 126. DOI: 10.2499/9780896295513

[6] Velásquez MA, Medina García G, Sánchez Cohen I, Klauidia Oleschko L, Ruiz Corral JA, Korvin G. Spatial variability of the Hurst exponent for the daily scale rainfall series in the state of Zacatecas Mexico. Journal of Applied Meteorology and Climatology. 2013;**52**:2771-2780. DOI: 10.1175/JAMC-D-13-0136.1

[7] González MG. Las zonas áridas y semiáridas de México, y su vegetación. Primera ed. SEMARNAT-INECC. Impreso en México; 2012. p. 194

[8] CNA. Programa Hidráulico de Gran Visión 2001-2025. Torreón, Coahuila: Gerencia Regional de las Cuencas Centrales del Norte; 2001

[9] CNA. Atlas Digital del Agua (2010). Contraste Regional Entre el Desarrollo y la Disponibilidad del Agua [Internet]. 2008. Available from: http://www.conagua.gob.mx/atlas/atlas.html?seccion=0<mapa=8 [Accessed: 16 October 2014]

[10] Díaz PG, Cortina CMF, Rodríguez PL. Potencial agro-productivo de cada hectárea de la superficie territorial del estado de Veracruz. Décima cuarta Reunión científica—tecnológica forestal y agropecuaria; 2001

[11] Barón WVS. Zonificación agroclimática para el establecimiento de sistemas silvopastoriles con Paulownia elongata [thesis]. Chapingo, México: Universidad Autónoma Chapingo; 2013

[12] FAO. Agua para la alimentación. Agua para la vida. Una evaluación exhaustiva de la gestión del agua en la agricultura. Washington, D.C.; 2007. p. 57 [Internet]. 2007. Available from: http:www.academia.edu/30158519/Agua_para_la_alimentación_Agua_para_la_vida [Accessed: 17 June 2015]

[13] Yáñez CLG, Pedroza SA, Martínez SM, Sánchez CI, Echavarría CJG, Velásquez VMA, López SA. Caracterización fisiográfica y regionalización de la Cuenca media río Nazas-Aguanaval. Revista Chapingo Serie Zonas Áridas. 2017 [in progress]

[14] Pedroza SA. El déficit hídrico en las plantas. Principios y técnicas de manejo. Bermejillo, Dgo: Universidad Autónoma Chapingo, Unidad Regional Universitaria de Zonas Áridas; 1995. 162 p

[15] INEGI (Instituto Nacional de Estadística, Geografía, e Informática). Descarga de mapas digitales de la República Mexicana [Internet]. 2015. Available from: http://www.inegi.org.mx/geo/contenidos/recnat/default.aspx [Accessed: 17 May 2016]

[16] ESRI. ArcGIS 10.1; also contains ESRI Data and Maps for ArcGIS. Redlands, CA; 2012

[17] Stretta EJ, Mosiño PA. Distribución de las zonas áridas de la república según un nuevo índice de aridez derivado del de Emberger. Ingeniería Hidráulica en México. 1963;**16**: 40-47

[18] SMN (Servicio Meteorológico Nacional). Normales climatológicas. México D.F.: Comisión Nacional del Agua; 2010

[19] Etchevers BJD, Gómez DJD, Monterroso RAI, Tinoco RJA. Formulación de indicadores para evaluar y monitorear la desertificación en México. Secretaría de Medio Ambiente y Recursos Naturales, Instituto de Ecología, Colegio de Posgraduados, Universidad Autónoma Chapingo; 2013. 155 p

[20] Velasco I, Reyes O. Caracterización hídrica del Estado de Sinaloa, mediante el índice de aridez y el régimen de aridez. Instituto Mexicano de Tecnología del Agua. Resumen Núm.028 del Congreso de Ciencias Ambientales. Universidad Nacional Autónoma del Estado de México. Mesa de Recursos Naturales [Internet]. 2007. Available from: www.uaemex.mx/Red_Ambientales/docs/congresos/.../RN028.doc [Accessed: 15 April 2015]

[21] Stankey G, Cole D, Lucas R, Petersen M, Frissell S. The limit of acceptable change, system for wilderness planning. General Technical Report INT-176. Ogden, UT, USA: U.S. Department of Agriculture. Forest Service. Intermountain Forest and Range Experiment Station; 1985. p. 37

[22] FAO. Secuestro de carbono en tierras áridas. Roma: Organización de las Naciones Unidas para la Agricultura y la Alimentación; 2007. p 138

[23] UACh. Actualización de la delimitación de las zonas áridas, semiáridas y subhúmedas de México a escala regional. Chapingo, México: Departamento de Suelos, Universidad Autónoma Chapingo; 2011

[24] Cruz Martínez A, Pedroza Sandoval A, Trejo Calzada R, Sánchez Cohen I, Samaniego Gaxiola JA, Cantú Brito JE. Alternatives and decision-making in natural resource management in degraded dryland areas. Revista Chapingo Serie Zonas Áridas. 2015;**14**(1):51-60. DOI: 10.5154/r. rchsza.2014.12.004

[25] Sánchez Cohen I, Estrada Avalos J, Cueto Wong J. Toma de decisiones en grupo para el manejo de los recursos naturales. Métodos de análisis y criterios de selección. Folleto Científico No. 24. Gómez Palacio, Dgo: INIFAP CENID RASPA; 2008. 73 p

[26] Heilman P, Stone J, Sánchez Cohen I, Macías Rodríguez H, Mann RS. Working smarter: Research and decision support systems in Mexican agriculture. In: Richardson C, Baez D, Tiscareño M, editors. Modeling and Remote Sensing Applied to Agriculture. U.S. and Mexico; 2005. pp. 211-236

[27] Cruz Martínez A, Pedroza Sandoval A, Trejo Calzada R, Sánchez Cohen I, Samaniego Gaxiola JA, Hernández Salgado R. Captación de agua de lluvia y retención de humedad edáfica en el establecimiento de buffel (Cenchrus ciliaris l). Revista Mexicana de Ciencias Pecuarias. 2016;**7**(2):159-172

[28] INIFAP. Red de Estaciones Agroclimáticas. Secretaría de Agricultura Ganadería y Pesca [Internet]. 2014. Available from: http://clima.inifap.gob.mx/lnmysr [Accessed: 14 May 2014]

[29] Idrobo H, Díaz AM, Ortiz JE. Comportamiento del hidrogel en suelos arenosos. Ingeniería de los recursos naturales y del ambiente. 2010;**19**:27-31

[30] Rojas GB, Ramírez M, Aguilera R, Prin JL, Torres C. Los hidrogeles poliméricos como potenciales reservorios de agua y su aplicación en la germinación de semillas de tomate en diferentes tipos de suelos. Revista Iberoamericana de Polímeros. 2006;**7**(3):199-210

[31] Beltrán LS, Loredo OC. Reconversión de áreas agrícolas marginales a praderas de pasto buffel. Folleto Técnico para Productores. INIFAP; 2002

[32] Burquez A. El zacate buffel: transformación ecológica y social. CONABIO. Biodiversitas. 2007;**74**:8-12

[33] Alcalá GC. Guía práctica para el establecimiento, manejo y utilización del zacate buffel. Patronato del Centro de Investigaciones Pecuarias del Estado de Sonora, A. C. [Internet]. 1995. Available from: http://www.patrocipes.org.mx/publicaciones/pastizales/p95009.php [Accessed: 15 April 2014]

[34] Gutiérrez CIJ, Sánchez CI, Cueto WJ, Trucios CR, Trejo CR, Flores HA. Efecto del polímero Aquasorb® en la capacidad de retención de humedad del suelo y su efecto en el rendimiento de la acelga (Beta vulgaris var. cycla). Revista Chapingo Serie Zonas Áridas. 2008;**7**(1):66-72

[35] Rivera HC, Baeza AA, Chavarriaga MW. Efecto de un retenedor de agua y dosis crecientes de fertilizantes foliares sobre la producción de tomate chonto y larga vida bajo cubierta plástica. Agroclear Agron. 2007;**15**:1-17

[36] Tezara WM, Driscoll SD, Lawlor DW. Water stress inhibits plant photosynthesis by decreasing coupling factor and ATP. Nature. 1999;**1401**:914-917

[37] Yáñez CLG, Pedroza SA, Martínez SM, Sánchez CI, Echavarría CJG, Velásquez VMA, López SA. Retención de humedad edáfica en la sobrevivencia y crecimiento de dos especies de pastos *Bouteloua curtipendula* [Michx.] Torr. y Chloris gayana Kunth en Durango, México. Revista Mexicana de Ciencias Pecuarias. 2017 [in progress]

[38] Carrillo SSM, Arredondo MT, Huber-Sannwald E, Flores RJ. Comparación en la germinación de semillas y crecimiento de plántulas entre gramíneas nativas y exóticas del pastizal semiárido. Revista Mexicana de Ciencias Pecuarias. 2009;**47**(3):299-312

[39] Esqueda CMH, Melgoza CA, Sosa CM, Carrilo RR, Jiménez CJ. Emergencia y sobrevivencia de gramíneas con diferentes secuencias de humedad-sequía en tres tipos de suelo. Revista Mexicana de Ciencias Pecuarias. 2012;**43**(1):101-115

[40] Barón CA, Barrera RIX, Boada ELF, Rodríguez NG. Evaluación de hidrogeles para aplicaciones agroforestales. Revista Ingeniería e Investigación. 2007;**27**:35-44

[41] Rehman R, Ahmad R, Safdar M. Effect of hydrogel on the performance of aerobic rice sown under different techniques. Plant, Soil and Environment. 2011;**57**:321-325

[42] Narjary B, Aggarwal P, Singh A, Chakraborty D, Singh R. Water availability in different soils in relation to hydrogel application. Geoderma. 2012;**187**:94-101

[43] Gómez MS, González DJR, Aguilar PD. Evaluación de híbridos experimentales de zacate buffel en suelo arcilloso del norte centro de México. Memorias del VI Congreso Internacional de Manejo de Pastizales. Durango, Dgo; 2015. 467 p

[44] González DJR, Gómez MS, López DA. El transplante garantiza establecer zacates forrajeros en suelos salinos y arcillosos. Memorias del VI Congreso Internacional de Manejo de Pastizales. Durango, Dgo; 2015. 464 p

[45] Nissen J, Ovando C. Efecto de un hidrogel humectado aplicado a las raices de nothofagus obliqua (MIRB.) OERST. Y Nothofagm dombeyi (MIRB.) OERST. durante su trasplante. Agro sur. 1999;**27**(2):48-58

[46] Ravisankar N, Balakrishnan M, Ambast SK, Srivastava RC, Bommayasamy N, Subramani T. Influence of irrigation and crop residue mulching on yield and water productivity of table purpose groundnut (Arachis hypogaea) in humid tropical island. Legume Research: An International Journal. 2014;**37**(2):195-200. DOI: 10.5958/j.0976-0571.37.2.029

[47] Kusmiyati F, Sumarsono S, Karno K, Pangestu E. Effect of mulch and mixed cropping grass-legume at saline soil on growth, forage yield and nutritional quality of guinea grass. Journal of the Indonesian Tropical Animal Agriculture. 2013;**38**(1):72-78

[48] Beggy HM, Fehmi JS. Effect of surface roughness and mulch on semi-arid revegetation success, soil chemistry and soil movement. Catena. 2016;**143**:215-220

Chapter 2

Production of Spineless Cactus in Brazilian Semiarid

Wilma Cristina Cavalcante dos Santos Sá,
Edson Mauro Santos, Juliana Silva de Oliveira and
Alexandre Fernandes Perazzo

Additional information is available at the end of the chapter

http://dx.doi.org/10.5772/intechopen.69614

Abstract

The term "spineless cactus" is used in Brazil to designate cultivars of *Opuntia ficus indica* Mill and *Nopalea cochenillifera* Salm Dyck. The spineless cactus was consolidated in Brazilian semiarid as a strategic fundamental food resource in several production livestock systems, constituting a plant with enormous productive potential. Thus, the spineless cactus has been widely cultivated and used for several decades, by enabling the animal feeding in critical periods of year because of its characteristics, morpho-anatomical and physiological (CAM), which makes it tolerant to long droughts, being a crop that presents high productivity in droughts conditions, when compared to other forages. Nevertheless, the spineless cactus is a crop relatively picky about soil and climate characteristics of region, presenting greater growth in fertile soils, as well as in regions where nighttime temperatures are cool and the air humidity is relatively high. Although the crop be adapted to long droughts periods, many times it's necessary to perform irrigation in its production system, mainly in regions of low rainfall, for to supply its water needs, thus ensuring productivity and survival of crop. Therefore, the knowledge of characteristics of plant, as well as of appropriate management techniques to crop, is essential for the good performance of spineless cactus.

Keywords: *opuntia ficus indica, nopalea cochenillifera*, productivity, adaptability, requirements

1. Introduction

The spineless cactus is a native cactus of semiarid regions of American continent, specifically from Mexico, being cultivated for forage and fruits production. In other regions of the world, beyond to be used as forage resource, the spineless cactus is cultivated for medicinal purposes, cosmetics, dyes, vegetable production, fruit production [1], fences and landscaping, and in some countries of Africa, the spineless cactus is a part of humans' diet.

However, the spineless cactus has been consolidated in arid and semiarid regions of the world as forage strategic in various production livestock systems [2] for being a culture adapted to soil and climate conditions, in addition to presenting high dry matter production per unit of area [3].

The date its introduction in Brazil remains obscure, having multiple versions in literature, and most of it are not based on more rigorous historiographical study [4]. Even so, there are reports in literature of your likely introduction in Brazil during the colonial period, being introduced in Rio de Janeiro by Portuguese, aiming to preclude the Spanish monopoly about the red dye Carmine produced in Mexico [5]. The pragmatism of that movement manifested itself especially during the administration of Marquis of Pombal, where the Portuguese Crown encouraged the colonies to produce natural products and the study, especially of Botany [6]. Around 1880, Herman Lundgren introduced in Pernambuco spineless cactus originating in Texas, where they were studied by the botanist Burbanks [7].

At first, the forage value of spineless cactus in Brazilian semiarid region was not recognized, although, in North Africa, the cultivation of varieties of *Opuntia* for fodder purposes was widespread in the late nineteenth century [7]. The spineless cactus only aroused interest as feed in Pernambuco State and Alagoas State in 1902 [5]. In early twentieth century, after the drought of 1932, order of government began to spread the spineless cactus [8], realizing that the little established plantations were insistently searched for cattle, goats and sheep that ate. Thus, the plant began to be used by animal breeders [6].

From the late 1950s, that really started the deeper character studies on the species, for to improve your use. Between 1979 and 1983, during the prolonged drought in Brazil's northeast, the spineless cactus won your space in semiarid scenario [9]. From this date, numerous studies have turned to this forage.

In recent years, the spineless cactus went back to being cultivated on a large scale by the creators of dairy cows [10] and it is estimated that today there are about 600,000 hectares of spineless cactus in Brazil's northeast [2], and a large part of these hectares concentrated in States of Pernambuco, Paraíba, Alagoas, Rio Grande do Norte and Bahia [6, 11].

2. The spineless cactus in Brazilian semiarid

The Brazil's northeast is the region which is the largest cultivation area of spineless cactus throughout world, with about 600,000 hectares, and the most commonly used cultivars are the Gigante, the Redonda and the Miúda, being that the choice has been determined by soil and climate conditions of planting sites. The Miúda cultivar is planted on a large scale in State of Alagoas, while in other northeast states (Pernambuco, Paraíba and some regions of Ceará and Rio Grande do Norte) it predominates the planting of cultivars of *Opuntia ficus indica* [2, 12, 13].

The Gigante cultivar (*Opuntia ficus indica* Mill) is a plant well developed with stem little branched, which gives an aspect upright and vertical growth little leafy. It possesses characteristics like its arborescent size with 3–5 m of height, broad crown, glabrous and 60–150 cm of width of stem. Its cladode weighs about 1 kg, showing up to 50 cm long, oval-elliptic or suboval form and matte green coloration. The flowers are hermaphrodite, of medium size, bright yellow coloration and petal that stays open at anthesis. The fruit is an ovoid berry, large, yellow, changing to purple when ripe. This cultivar is considered the most productive and more resistant to drought regions; however, it is less palatable for animals and smaller nutritional value [14–16] (**Figure 1**).

The Redonda cultivar (*Opuntia* sp.), originated of Gigante cultivar, has medium size and stem many branched laterally, thereby reducing the vertical growth. Its cladode weighs about 1.8 kg, owning nearly 40 cm long, round and ovoid form. It presents great yields of a material more tender and palatable than the Gigante cultivar. Its lateral growth hinders the intercropping with annual crops, and thus, has been less common the planting with this cultivar [14–16] (**Figure 2**).

Figure 1. Morphological aspect of Gigante cultivar. Photo: Antônio Carlos Alves.

Figure 2. Morphological aspect of Redonda cultivar. Photo: Renaldo Araújo.

On the other hand, the Miúda cultivar (*Nopalea cochenillifera* Salm Dyck) has small size and stem quite branched. Its cladode weighs about 350 g, has almost 25 cm long, sharply obovate form (apex wider than the base) and intense bright green coloration. The flowers are red, and your petal remains half closed during the cycle. The fruit is a purple berry. Comparing with the previous two cultivars, this is the most nutritious and appreciated by animals (palatable), but offers less resistance to drought. It is a most demanding cultivar in soil fertility [14–16]. It is more demanding in humidity and temperature cooler to night when compared to the other cultivars [16, 17] (**Figure 3**).

The spineless cactus (*Opuntia* and *Nopalea*) is species that stand out in Brazilian semiarid region, having contributed significantly to livestock feed in prolonged droughts, since their anatomical and physiological features allow your productivity. In the three plants, the cladodes are covered by a cuticle which controls the evaporation, allowing the storage of water until the level of 90–93% [18].

In general, we can say that the cultivars of *Opuntia ficus indica* have shown more rustic when compared to Miúda cultivar, due to larger tolerance to intense droughts and pest attack *Diaspis echinocacti*, commonly known as "scale cochineal" [13]. However, the Miúda cultivar presents resistance to "carmine cochineal" (*Dactylopius opuntiae*), which is currently the main plague of spineless cactus culture in Brazil's northeast, and for this reason, there is a tendency to increase the planting area with this cultivar [7].

Regarding productivity [12], the Miúda cultivar has shown to be smaller than the Gigante and Redonda cultivars; however, when this production is considered in terms of dry matter, the results are equivalent, since the cultivar Miúda has higher dry matter content than the cultivars of genus *Opuntia*. Although it is considered as an excellent energy source (rich in non-fibrous carbohydrate, important source of energy and TDN) [2], the spineless cactus presents insufficient levels of neutral detergent fiber and crude protein for proper animal performance when provided as bulky food alone; therefore, the association with bulky foods of highly

Figure 3. Morphological aspect of Miúda cultivar. Photo: Agefran Costa.

effective fiber content and non-protein nitrogen sources and/or true protein is required [19] (**Table 1**).

Because of its low dry matter content and high-water content, the use of spineless cactus in isolation in animal nutrition is not recommended and should be commonly used to compose the diet, replacing partially traditional forage [16]. Best result is achieved in fiber consumption by sheep when the spineless cactus was mixed to a diet of hay and concentrate [23]. Pessoa et al. [24] investigated the effects of different food strategies in spineless cactus-based diets, associated with sorghum silage and concentrated on the performance of dairy cows, and stated that the strategy of mixing the ingredients completely provided balance in the supply of nutrients for animals (protein, energy, effective fiber, minerals, etc.), because it made possible the decrease in the selection of ingredients, providing suitable relationship bulky/concentrate on diet and, consequently, the ruminal environment health, with gains in productivity.

Araújo et al. [25] evaluated the effect of use of two cultivars of spineless cactus (Gigante and Miúda) with and without the addition of maize in diet of lactating cows, noting that the

	Spineless cactus's cultivars		
Nutrients	***Opuntia ficus indica* Mill**	***Opuntia* sp.**	***Nopalea cochenillifera* Salm Dyck**
Dry matter (% as fed)	10.2	11.0	15.4
Organic matter (% DM)	89.8	89.1	93.0
Crude protein (% DM)	5.3	5.2	3.5
Neutral detergent fiber (% DM)	26.0	26.2	25.8
Acid detergent fiber (% DM)	22.4	22.2	23.0
Non-fibrous carbohydrate (% DM)	55.6	-	71.2
Total carbohydrate (% DM)	81.9	81.2	87.8
Total digestible nutrients (% DM)	64.3	-	-
Ether extract (% DM)	1.98	1.78	1.71
Mineral matter (% DM)	11.2	11.2	7.0
Crude fiber (% DM)	12.3	8.7	7.17
Non-nitrogenous extractive (% DM)	70.3	72.8	78.0
Calcium (% DM)	2.1	2.9	3.8
Phosphorus (% DM)	0.1	0.1	0.2
Potassium (% DM)	2.1	2.5	1.5

Table 1. Nutritional composition of different spineless cactus's cultivars. Source: Adapted from Refs. [20, 21, 22]

consumption of dry matter was not influenced by cultivars of spineless cactus studied, and, however, found higher consumption for diets with corn, which had higher dry matter content than those without corn, factor that possibly determined this difference. Wanderley et al. [26] evaluated the consumption of lactating cows fed with feed containing levels of spineless cactus + sorghum silage + concentrate, noting increase in dry matter intake, that according to authors was due to supply of food in form of complete feed, which provided, throughout the day, better supply of nutrients, favoring and conforming the ruminal fermentation, mainly to concentration of volatile fatty acids. The authors stressed the importance of animals has not been presented metabolic disorders, such as diarrhea, when spineless cactus was supplied under this food strategy, in association with fiber-rich food sources.

However, it is important to note that the high-water content of spineless cactus is an indirect way of promoting greater water consumption in diet [16], an important factor for the creation of animals in arid and semiarid regions [27], because in a region where water is scarce and often of bad quality, this characteristic must be framed among the positive aspects of forage [28].

In arid and semiarid regions, the spineless cactus has been the basis of ruminant feed because it is a culture adapted to soil and climate conditions, in addition to presenting high dry matter production per unit of area [3].

Recently, studies have been developed, seeking the intensification and efficiency in the use of spineless cactus to reduce the time and labor costs for harvesting and daily supply of animals. Thinking about this, the research has been focused on production of silage, since it would allow the maximization of the use of this forage, as well as improve operational logistics in supplying food diary to animals. In this way, the spineless cactus ensilage would allow harvest of all the plantation, standardizing and increasing regrowth capacity and, consequently, the productivity, beyond to reduce labor with harvest and periodic supply, throughout the dry season.

Although spineless cactus presented some unfavorable characteristics to ensilage, such as low dry matter content and highly soluble carbohydrate concentration, favoring growth of undesirable microorganisms, it has features that distinguish it from other foragers. The mucilage of spineless cactus is constituted by hydrocolloids which are distributed throughout the plant and have the property of water absorption [29, 30]. The hydrocolloids are compounds formed by highly hydrophilic polysaccharides that minimize the movement of the water, providing the increased viscosity of material and thus the formation of mucilage.

It should be noted that spineless cactus has bioactive compounds, such as organic acids (malic, citric, oxalic, malonic, succinic and tartaric acid) found in their cladodes [1]. The presence of these substances buffers can control the growth of yeasts through buffering of ensiled mass, directing the fermentation to produce lactic acid, thereby minimizing losses during ensilage [31].

Beyond these characteristics presented by spineless cactus, the silage additives are added to forage for to correct characteristics unfavorable during the ensiling process.

About the exposed, studies [32] showed the efficiency of spineless cactus for ensilage. This author evaluated the potential of spineless cactus for ensilage without additives or additive with wheat bran and urea, noting that pH values varied between 3.7 and 4.2, values considered ideal for well-fermented silages [33]. Really, they found lactic acid production close to 100 g/kg in silages with or without additives—content considered normal for fermented silages by acid lactic acid bacteria [33]. It should be noted that spineless cactus used in this work presented 12% of dry matter, soluble carbohydrates content of 120 g/kg of dry matter and a buffer capacity of 22 mEq/100 g DM. The combination of these three characteristics can result in a high fermentative capacity, without, however, trigger alcoholic fermentations.

Recently, Sá et al. [34] evaluated silages of five complete feed based on spineless cactus in three opening times (7, 15 and 60 days), and noted that all silages showed pH values indicative of normal fermentation, no difference between the feed in each open, decreasing significantly to 60 days, with an average of 3.98. In this study, the concentrations of lactic acid of feed significantly increased to 60 days, reaching 17.34% based on DM.

Brito et al. [35] evaluated the spineless cactus silage with chemical additives (2% urea based on DM) and microbial (*Lactobacillus buchneri*), as well as the association of both (2% urea + *Lactobacillus buchneri*) in four opening times (7, 15, 60 and 120 days) and observed that all silages showed values of pH considered suitable for silage well fermented, around 4.0. In these silages, lactic acid levels increased significantly from 60 days, reaching 8.49% based on DM to 120 days.

However, despite the excellent quality of spineless cactus silage, the performance assessment studies of animals consuming such silage are virtually nonexistent in Brazilian semiarid region. Nevertheless, unpublished data on performance evaluation of sheep getting complete feed silage based on spineless cactus showed satisfactory results. Therefore, more studies are needed to behold the performance of animals consuming spineless cactus silage in Brazil's semiarid region.

3. Adaptive characteristics of spineless cactus

The spineless cactus is considered a xerophyte plant due to the fact that its adaptive features allow your survival in hot and dry environments.

Xerophytic plants are characterized by structural modifications (physiological and morphological) that help these plant species survive in the more complicated climatic conditions that are hot and dry climates, which often does not have the ideal amount of water to grow a plant. In Brazilian semiarid region, especially in drought periods, water is a rare item, including for the human beings themselves. So, the xerophytic plants needed to develop mechanisms to make them support these adverse conditions and they could survive. Among the mechanisms and adaptations, morfoanatômicas developed by plants xerophytic are:

- Dense nerves; epidermal cells small; bristle coating; external walls of the epidermis thickened; very developed sclerenchyma; thick cuticle; cutinized layers; presence of wax, tannins, volatile oils, resins, mucilage and various layers of palisades [36, 37];

- Trichomes and many small stomata per unit of surface, inside of crypts formed by cutine layers on the epidermis [7];
- Small-size leaves that are waxy and, often, the leaves these plants are modified to thorns, as adaptation, that cause smaller loss of water, making the plant survive any longer;
- Stems and roots that can store water for the vital needs of the plants; strong roots that grow up and enter the soil to reach the underground water sheets [38].

Another adaptive mechanism of xerophytic plants is the ability to maintain high-water potential in the tissues, which is achieved by the absorption of water or decreasing water loss by transpiration. For maintenance of the water absorption, the plant can present a deepening or comprehensiveness of the root system, increased hydraulic conductivity and osmoregulation in the roots. And for the reduction of water loss by transpiration, the plant can promote the reduction of epidermis conductance through the thickening of cuticle, reducing the amount of radiation absorbed by production of bristle and wax, and reduced leaf area and stomata [37, 39, 40].

Another very important aspect of xerophytic plants when subjected to water stress is the osmotic adjustment, in other words, active fotossintetizados product buildup inside the cell [36, 37], which are used to promote the development of adaptive features of plant.

Unlike other xerophytic plants, spineless cactus presents a shallow root system and distributed horizontally, fleshy that exploring almost the entire surface of the soil (10–20 cm), with high-water absorption capacity of the light rain and even the dew, featuring an advantage in places of low rainfall [37, 41]. The distribution of spineless cactus roots may depend on ground conditions. Under favorable conditions of soil, moisture develops an elongated root. On the other hand, under dry conditions develop lateral fleshy roots from the main root to thus absorb water at shallow levels [42].

The root system of spineless cactus is very complex, and it can have four types of roots [42]:

The structural roots, formed by a primary with little fibrous roots skeleton of 20–30 cm in length, forming quickly a periderm, but keeping many latent and active gems, distributed from the base until the apical region without a regular pattern of distribution. When the structural roots remain dry for a while and suddenly are moistened, in a few hours if restarts the formation of absorbent roots that respond quickly to moisture.

The absorbing roots form within few hours after the side buds respond to moisture and are called "rain roots." These roots die as soon as the soil dries.

The spur roots are formed as the most voluminous mass of roots and can be short, thick and fleshy, with many fine bristle roots, and long, like the system of absorbent roots.

The roots of areolas develop when the areolas are in contact with the ground. At the beginning of its development are thick and without bristle and have a kalyptra with the cells of the epidermis forming appendages like bracts. The growth of young roots is very fast, and they become soft with a shell of three to four cells thick and are covered with many bristle roots. Over time, all roots that originate from areolas form a real root system.

The fine roots (<1 mm) are considered as the main in processes of absorption of water and nutrients for plant, being observed wide variation in your distribution in the soil profile, depending on the genotype and sampling period [41].

In addition to these features, the physiology of spineless cactus is characterized by the photosynthetic process named Crassulacean Acid Metabolism (CAM). The CAM metabolism allows plants to improve efficiency in the use of water. Typically, a CAM plant loses 50–100 g of water for each gram of CO_2 obtained, whereas plants with metabolism C_3 and C_4 lose 400–500 and 250–300 g, respectively. Thus, CAM plants have a competitive advantage in dry environments [43].

A key feature of CAM plants is your juiciness due to its thick cladodes and large vacuoles filled with water in the photosynthetic cells, as well as of several layers of cells' water storage. The mature cladodes of spineless cactus usually have 1–5 cm thickness, and most of it is a whitish water-retentive tissue. The greenish chlorenchyma, which contains chlorophyll and where occurs photosynthesis, has a layer of 2–5 mm thickness on each side of cladode; it consists of 15–40 layers of compact cells. The water storage parenchyma also has compact layers of cells, slightly larger than the chlorenchyma. During drought, the water is preferentially lost from the parenchyma, allowing the chlorenchyma to remain well hydrated and allowing the continuity of photosynthesis [44].

Plants of CAM metabolism, unlike other plants of C_3 and C_4 metabolism, open their stomata at night and close during the day, which means the capture of atmospheric CO_2 takes place in the dark. This is considered a mechanism for adaptation of these plants to arid and semiarid regions, to minimize water loss. The clamping mechanism of CO_2 in these plants is very like the mechanism of C_4 plants; however, in CAM plants, the fixation of CO_2 occurs two-way [Rubisco and phosphoenolpyruvate (PEP) carboxylase], being separated in both time and spatially. Initially, the CO_2 is captured at night, via PEP carboxylase enzyme in cytosol, using the phosphoenolpyruvate (PEP) as acceptor and forming oxaloacetate which is then reduced to malate. The malate is stocked in large vacuoles, anatomical characteristic typical of leaf cells of CAM plants, acidifying them. The next day, with the stomata closed, the malate is transported to the chloroplast and decarboxylated by the enzyme NADP-malic to pyruvate and CO_2. Since the stomata are closed, the CO_2 released internally cannot escape, being refixed via Calvin-Benson cycle, by Rubisco, and converted to carbohydrates (**Figure 4**). The high inside concentration of CO_2 favors activity carboxylative of Rubisco [39, 43].

The key to water conservation by CAM metabolism plants is the opening of stomata at night, resulting in less water loss. The water loss from a CAM plant is much smaller than that of other species (plants C_3 and C_4) due to the lower proportion of surface area open to the atmosphere. In addition, the cooler temperature at night makes you reduce the difference of the water vapor content between the plants and the air around them (**Figure 5A**). Thus, during a period of 24 h, the spineless cactus can transpire 11.3 Moles (203 g) of water per m^2 of surface, while plants C_3 and C_4 can lose about 4.7 and 2.9 times more, respectively [44].

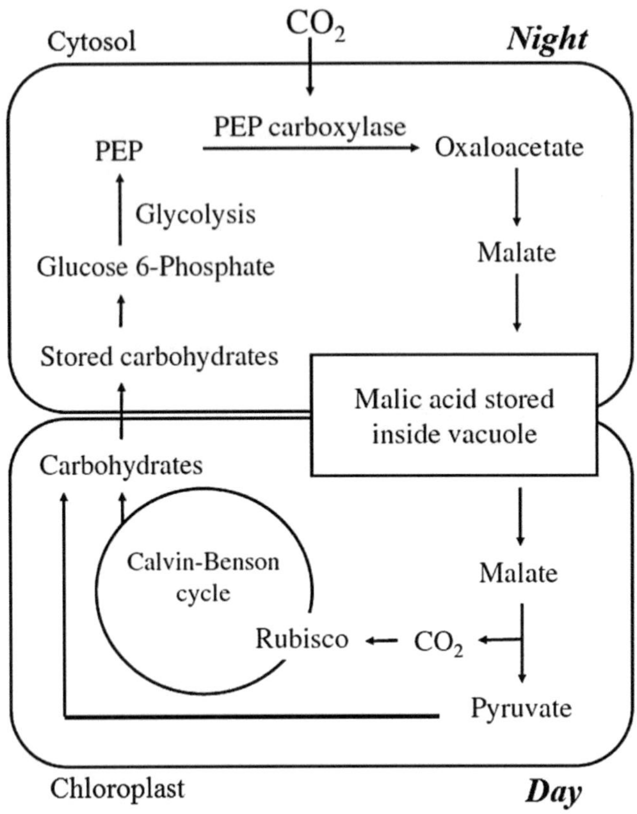

Figure 4. Crassulacean acid metabolism schema (CAM).

Previous studies [44] discuss evaluation of plants with different types of photosynthetic metabolism (C_3, C_4 and CAM), irrigated and fertilized without shade on bright days with maximum temperatures of 30–35°C and minimum night temperatures of 15–20°C; Nobel [44] noted that the capture of atmospheric CO_2 per hour between representatives of the three types of photosynthetic system is like the daily loss of water per hour. According to the author, the net speed of atmospheric absorption of CO_2 by nearly horizontal sheets of plants C_3 and C_4 gradually increases during the morning, as the sun rises, and reduces similarly in the afternoon, as the incidence of light on the leaves decreases, with a near zero catch at dawn (**Figure 5B**).

Many plants C_3 tend to a partial closure of stomata close to noon, which results in the reduction of water loss, but also in reducing atmospheric capture of CO_2. The maximum speed of atmospheric capture of CO_2 tends to be greater in cultures C_4 and smaller in CAM species, although its speeds of absorption may be significant during the night. In addition, CAM plants well irrigated usually absorb some CO_2 in the morning and in the late afternoon, while the plants C_3 and C_4 do not absorb nothing during the night [43, 44].

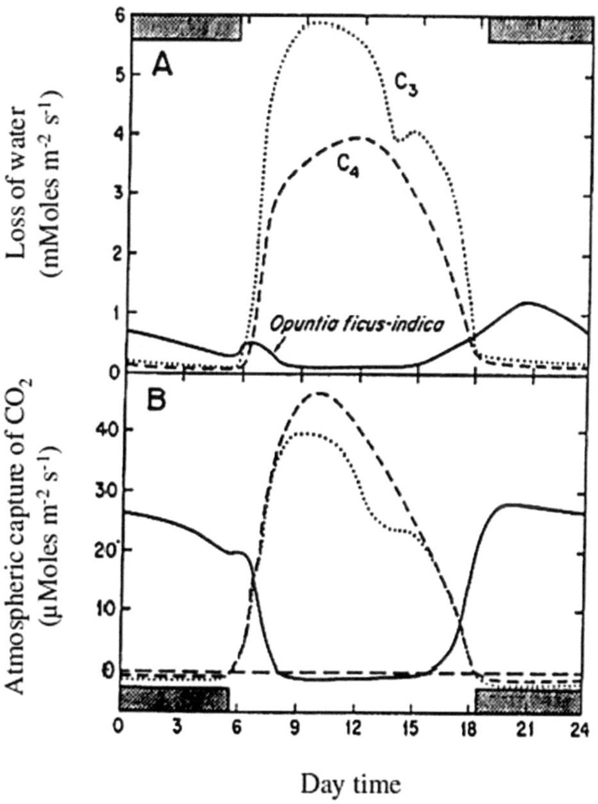

Figure 5. Daily loss of water (A) and atmospheric capture of CO_2 (B) for plants of type C_3, C_4 and for the species *Opuntia ficus indica* (CAM) (shaded area indicates the night). Source: [44].

4. Soil and climate requirements of spineless cactus in Brazilian semiarid

The semiarid region of Brazil's northeast is characterized by irregular rainfall, with rainfall between 300 and 500 mm/year, concentrated on a few months of year, consequently leading to long periods of drought. However, the spineless cactus is a plant adapted that has a good development in regions with little rainfall. Nevertheless, information about air and soil humidity, average temperature of day and night are crucial for production [16].

Climatic conditions exert a strong influence on growth and development of this plant [14]. Knowledge of phenology and the characteristics of cultures, when associated with the climatic conditions of their regions of origin and commercial dispersion, allows to establish the limits of climate requirement of species [45]. Thus, Souza et al. [46] have elaborated an agricultural zoning, using as essential tools, the information of phenology and the characteristics of the culture associated with the climatic conditions of the regions of origin and commercial dispersion of spineless cactus (**Table 2**).

	Aptitude		
Climate parameter	**Ideal**	**Restricted**	**Inadequate**
Average temperature (°C)	16.1 ≤ AverT ≤ 25.4	AverT < 16.1 and AverT > 25.4	-
Maximum temperature (°C)	28.5 ≤ MaxT ≤ 31.5	MaxT < 28.5 and MaxT 31.5	-
Minimum temperature (°C)	8.6 ≤ MinT ≤ 20.4	MinT < 8.6 and MinT > 20.4	-
Thermal amplitude (°C)	10.0 ≤ TA ≤ 17.2	TA < 10.0 and TA > 17.2	-
Precipitation (mm)	368.4 ≤ P ≤ 812.4	812.4 ≤ P ≤ 1089.9 and P < 368.4	P > 1089.9
Moisture index (-)	−65.6 ≤ MI ≤ −31.8	−31.8 ≤ P ≤ −7.7 and MI < −65.6	MI > 7.7

Source: Adapted from Ref. [46].

Table 2. Climatic indicators of agricultural zoning of spineless cactus.

Moura et al. [47] performed the agricultural zoning of spineless cactus for Pernambuco State based on climatic indicators presented by Souza et al. [46], as well as rainfall precipitation and climate data belonging to the Northeast Development Superintendence (SUDENE) and the National Institute of Meteorology (INMET), respectively. The authors observed that with respect to thermal amplitude, the favorable conditions for the cultivation of spineless cactus cover virtually the entire state. However, there may be thermal limitation for the cultivation of species in coastline, because of decreased of thermal amplitude in this territorial range. Also, it is observed that with respect to moisture index, approximately half of Pernambucan territory offers favorable conditions for the cultivation of spineless cactus, covering rural regions and part of Hinterland of State. However, when approached of arid region of São Francisco, it was verified restriction to cultivation, since this region has low values of MI, resulting, mainly from low levels of rain precipitation and greater evaporative demand, that condition the reduction in water content of soil. In contrast, the transition regions and the coastline were restricted and inadequate, respectively, which is associated with the excess rains resulting in increased moisture index. Finally, the results show that, under the climatic point of view, about 42.3% of state present conditions suitable for the cultivation of spineless cactus, while 54.4% are of territorial scope feature restrictions. In these areas, spineless cactus cultivation can be carried out; however, there may be restrictions regarding thermal amplitude or moisture index, which can result in a reduction in productivity.

For the cultivation of spineless cactus in low-risk climate conditions, in State of Rio Grande do Norte was elaborated an agricultural zoning, establishing the following criteria: average annual temperature (16–27°C), maximum temperature (28.5–33°C), minimum temperature (8.5–22°C) and average annual precipitation (360–800 mm/year). The municipalities that presented in at least 20% of its areas, thermal and water conditions within of the criteria established in at least 80% of the evaluated years were considered suitable for the cultivation of spineless cactus [48].

Bezerra et al. [49] determined the agricultural zoning of spineless cactus' cultivars for the municipality of Paraíba based on climatic indicators presented by Souza et al. [46] and in each of the meteorological stations in the State of Paraíba. The authors concluded that the mesoregion of Borborema and part of west-center mesoregion of region Agreste are the areas that present the most favorable climatic conditions for the cultivation of spineless cactus in state, in accordance with the ideal aptitude observed in **Table 2**. On the other hand, the coastline region of Paraiba and swamp of altitude around the municipality of Areia present the most unfavorable climatic conditions for the cultivation of spineless cactus. The mesoregion of Agreste and the arid region feature restrictions to precipitation and/or temperature.

Under conditions of excessive moisture in the soil accumulates water in quantities exceeding the transpiration capacity of plant, which facilitates the occurrence of rot, tipping and only then becomes highly vulnerable to diseases, especially those caused by fungi [49].

Accordingly, [14] the good yield of crops in semiarid northeast Brazil is associated with fact they need far less water than other conventional crops. The spineless cactus uses 100–200 kg of water to produce 1 kg of dry matter and produces well in areas with annual precipitation of up to 750 mm. It grows best where the average relative humidity of the air is above 40%, and day and night temperatures oscillate around 25 and 15°C. In some semiarid regions, low relative humidity and high nighttime temperatures are the main factors for the lower productivity or even death of plants [22].

Spineless cactus growth is favored in the higher altitudes, due to the reduction in air temperature and increasing relative humidity at night (55–60%) [13].

The spineless cactus is a culture relatively picky about physical and chemical characteristics of soil, showing greater growth in fertile soils. Therefore, if they are fertile, spineless cactus cultivation can be realized in areas of texture sandy to clay, but more often recommended the clay-sandy soils. In addition, fertility is also important that soil is well drained, since very moist soils do not lend themselves to the cultivation of spineless cactus [22], because it does not tolerate disabled drainage areas. The cultivation is also impossible in regions whose annual rainfall exceeds 1100 mm [46]. In addition, the spineless cactus does not tolerate high levels of salts [7]; therefore, it is not recommended to your cultivation in saline soils.

The spineless cactus is found in a wide range of soils, where the soil pH range is subacids to subalkalines, showing a good adaptation of species. Soils with 60–70 cm depth are good for the development of shallow root system of culture. However, soils with little drainage capacity, shallow groundwater and/or surface layer waterproof should not be regarded as adequate. The clay content must not exceed 15–20%, to avoid putrefaction of the roots [50].

5. Productivity of spineless cactus in Brazilian semiarid

5.1. Nonirrigated soil

Forage production in dry soil conditions means that the crop is cultivated without irrigation in regions where annual rainfall can be less than 500 mm. The cultivation will depend, in addition to precipitation, on specific techniques that allow an efficient use of the limited soil moisture. However, the spineless cactus is a plant which features high productivity in non-irrigated conditions, compared to other fodder, especially when subjected to appropriate agronomic practices and when used plant with high production potential, and being able, the production of dry matter varies from 12 to 47 tons every 2 years [25].

This productivity of spineless cactus can be observed in the study by Silva et al. [51] with *Opuntia ficus indica* and *Nopalea cochenillifera,* fertilized (130 kg N/ha/2 years) in drought conditions. The authors checked most green biomass (163.0 t MV/ha) for *Opuntia ficus indica,* differing of *Nopalea cochenillifera* (124.3 FM tons/ha). However, when this productivity was considered in terms of dry biomass, the incomes were equivalent, showing an average of 12.6 tons DM/ha/2 years.

Almeida et al. [52] evaluated the productive performance of *Opuntia ficus indica* and *Nopalea cochenillifera,* subjected to organic fertilization treatments (30 tons manure/ha), chemistry (100 kg P/ha and 300 kg N/ha) and the association of both in dense planting (1.0 × 0.25 m) in Semiarid Bahia. The data showed that, regardless of treatments, the fresh and dry biomass productions were equivalent among the species. However, when compared individually, the biggest productions were observed when there was association of organic fertilizer with the chemical.

Silva et al. [53] evaluated the dry matter production of spineless cactus cultivated under different types of chemical fertilizer (150 kg P/ha; 200 kg N/ha + 150 kg P/ha; and 200 kg N/ha + 150 kg P/ha + 100 kg K/ha) and spacing (1.00 × 0.50 m; 2.00 × 0.25 × 1.00 × 3.00 and 0.25 m), to 620 days after planting. The average productivity of dry matter was 17.1 mg/ha. The plants under 1.00 × 0.50 spacing with NPK, NP and P produced more dry matter than plants without fertilization. In spacing 2.0 × 0.25 m and 3.0 × 1.0 × 0.25 m dry matter production was similar for different fertilization.

The spineless cactus extracts large amounts of nutrients from soil. Considering an average annual productivity of 20 tons DM/ha, this plant extracts, approximately 180 kg of N, 32 kg of P, 516 kg of K and 470 kg of Ca per hectare. Considering an average productivity of 40 tons biennial DM/ha and average levels in DM of N, P, K and Ca as being of 0.9%, 0.16%, 2.58% and 2.35%, respectively, the spineless cactus extracts about of 360 kg of N, 64 kg P, 1032 kg of K and 940 kg of Ca per hectare every 2 years, without considering the other macros and micronutrients [54].

Dubeux et al. [55] observed influence of population of plants in spineless cactus productivity in several municipalities in semiarid region of State of Pernambuco. Dry matter production varied from 6 to 17 tons/ha in density of 5000 plants/ha and from 17.8 to 33.7 tons/ha in density 40,000 plants/ha, when spaced 2.00 m × 1.00 m and 1.00 m × 0.25 m, respectively. When assessing the spineless cactus growth in four spaces (1.00 m × 1.00 m; 1.00 m × 0.50 m; 2.00 × 1.00 m; 2.00 m × 0.50 m), Ramos et al. [56] concluded that the spacing influenced the production of biomass per area and that efficiency of use of rain by spineless cactus is incremented with higher population densities, being the best results observed in the spacing of 1.00 m × 0.50 m, resulting in a greater quantity of forage produced per area and per unit of rain. According to Ref. [22], the spineless cactus dense cultivation, with up to 40,000 plants/ ha, has been used in the Brazilian semiarid region, resulting in high productivity (320 tons FM/ha).

5.1.1. Management of spineless cactus in nonirrigated soil

Choice of species or cultivar—In Brazilian semiarid region predominate three cultivars of spineless cactus, of which two belong to the species *Opuntia ficus indica* Mill (Gigante and Redonda) and one belongs to the species *Nopalea cochenillifera* Salm Dyck (Miúda). The Gigante and Redonda cultivars have shown greater rusticity due to its resistance to drought, when compared to Miúda cultivar, not being recommended its use in drought conditions. Therefore, in choice of species, one should opt for the more adaptable to region to be cultivated. On the other hand, the Miúda cultivar has larger palatability in relation to others, though more demanding on soil and lower nighttime temperature [17, 57].

Planting area—Contrary to popular belief, the spineless cactus has requirements for the physical-chemical characteristics of soil. For your cultivation can be indicated the soil of sandy to clay texture, being most recommended mixed texture soils (clay sandy). The most fertile soil of property for planting is recommended, preferably deep and free of acidity, salinity and stones. The soil must be well prepared and mainly already corrected. It is very important that the soil has good drainage. In practice, such as the planting of spineless cactus is normally in final third of the drought period, and the mechanized soil tillage, mainly in conventional ways, can generate many clods of land, due to low soil moisture in that period. The clay terrain is the most conducive to form clods. Therefore, more care should be spent on fixing the seedlings in grooves of planting, if there are many clods [57].

Choice of cladodes and forms of planting—The cladodes must be obtained from young plants, preferably the most productive, stain-free, clinical signs of disease and pest-free (especially the cochineal). Must make planting of cladodes of good development, preferably located in the middle of plant. The cladodes should be cut and separated at the junction of cladodes, with the aid of a knife sharp and clean, to avoid possible contamination. The cladodes should be stored in shade for a period of about 10–15 days for healing of wound caused by cutting. In Brazilian semiarid region, many forms of spineless cactus planting are found: cladodes in vertical position; positioning to 45° of inclination; and planting with overlapping cladodes, referring to a deck of cards open, bilateral alignment in groove referencing aligned domino pieces, among others. However, regardless of the form of planting, one must prioritize one east-west orientation to maximize uptake of solar radiation [17, 57].

Fertilizing—As for the fertilization of establishment has been studied the addition of organic fertilizers, minerals and the joint addition of these, obviously depending on factors, such as level of soil fertility, availability of financial resources, among others. If to add manure in groove, put a layer of land on the manure or spread the manure between planting lines avoiding the contact with the basis of plants. These measures ensure the reduction of plant mortality by rot of base cladode. It is valid to note that in forage use, the spineless cactus extracts considerably some specific soil nutrients (for 10 tons DM/ha/year: 90 kg N/ha, 16 kg P/ha, 258 kg K/ha and 235 kg Ca/ha), which need to be restored [57].

Planting spacing—The planting spacing to be used varies according to soil fertility, amount of rainfall, size of property, forage need and purpose, among other factors [57]. However, it should be chosen according to the preferences and the availability of capital from the producer [58].

The practice of dense cultivation makes it possible to achieve greater forage production by area; however, the costs of establishment of plantation also are larger, and the cultural practices become more difficult and do not allow cultivation with other crops [59]. It is worth mentioning that spineless cactus extracts large amounts of nutrients from the soil. By adopting a system of dense planting, there will be greater extraction of nutrients from the soil, so it must have greater care with fertilizing, because it can cause yellowing of the cladodes by nutrient deficiency [60]. Moreover, it can affect the light interception and photosynthetic efficiency, influencing on the development and productivity of the plant. In Brazilian semiarid region, usually if it adopts the spacing of 1 m between rows and 0.25 m between cladodes, while in other countries it used 3 m or more, which facilitates the mechanization. Thus, less dense plantations facilitate cultural practices with animal traction, important for family agriculture [61].

Care with crops—The control of invasive plants is of fundamental importance in cultivation of spineless cactus. In addition to competition for light, competition from invasive plants for water and nutrients, due to shallow root system of spineless cactus, reduces the productivity of this crop and increases the risk of fire [12].

Pest and disease control—About the pests and diseases, in Brazilian semiarid region, although there are records of diseases, the problems are small and localized [12]. With respect to pests, scale cochineal (*Diaspis echinocacti*) and carmine cochineal (*Dactylopius opuntiae*) are the main today. In relation to first, the biological control by predator insect (*Coccinella septempunctata*) known as "Joaninha" has been shown to be efficient, but the chemical control with mineral oil is also recommended when massive infestation occurs. In relation to second, the use of resistant cultivars, like the Miúda cultivar, has been shown to be very efficient [7].

Harvest—Usually, spineless cactus harvesting is made every two years. However, the cut-off frequency can vary depending on the need of producer and of the climatic conditions. Nevertheless, as well as for other forage, there is the need to preserve a residual cladode area to promote vigorous regrowth and increased longevity of plant [7, 12].

5.2. Irrigated soil

Despite all morpho-anatomical and physiological adaptability, growth and development of spineless cactus varies with the weather conditions, where often necessary irrigation events in its production system so that it can meet its water need [62].

It very common to irrigate it in areas with long periods of drought, where the spineless cactus is used for fruit production and human food (Mexico, Chile, Italy and Israel) [44, 63] for to supply its water need, especially in periods of low rainfall levels to ensure productivity and survival of crop [64].

The water deficit in soil negatively influences on growth and development of plants, since it reduces your water potential, resulting in loss of turgidity, closing of stomata, reducing growth and, consequently, reducing the final output [65].

In this way, the knowledge of how spineless cactus responds to different levels of water availability is considered as indispensable for the establishment of management strategies, which

http://dx.doi.org/10.5772/intechopen.69614

are aimed at better use of water reserves in soil by crop. Information, such as these are fundamental to the management of spineless cactus, in search of more efficient use of available water, considering that this is a very cultivated forage in areas of low water availability [62].

In State of Rio Grande do Norte was conducted a study with spineless cactus dense (50,000 plants/ha), fertilized (organic, 50 tons manure/ha; and chemistry, 500 kg superphosphate/ha in foundation and 225 kg of nitrogen/ha/year) and irrigated (7.5 liters per linear meter every 10 days, 3.75 mm) in municipalities of Apodi, Cruzeta and Pedro Avelino for to improve the performance of spineless cactus in these regions of State where it suffers severe wilting due to inadequate climatic conditions (high temperatures and low relative humidity) (**Table 3**). The published results proved the effectiveness of irrigation of salvation as enabling technology of spineless cactus production in semiarid Brazilian, where traditionally were not obtained productions on dryland cultivation system [66].

In general, researches developed by EMPARN [66] with spineless cactus irrigated and dense (50,000 plants/ha) achieved productivity average of 250–350 tons FM/ha in cuts with annual frequency. Dry matter yields are variable and dependent on the concentration of crop dry matter.

In Rio Grande do Norte, the first studies with spineless cactus cultivation under irrigation were performed by Wanderley in 1996, in municipalities of Lajes, Angicos e Pedro Avelino. After testing several alternatives, he defined a system with use of high densities of planting with 50,000–100,000 plants/ha and drip irrigation in simple rows with low intensity, 5 liters of water/linear meter (2.5 mm) every 15 days (5 mm/month), as well as organic and chemical fertilization. Even when it comes to empirical data, high productivity was obtained in a region where the spineless cactus had never previously succeeded [63].

It is also important to point out that those were the yields obtained by EMPARN in its experiments, which is not to say that yields larger or smaller cannot be obtained. Indeed, Queiroz et al. [67] evaluated the effect of application of different irrigation blades (976, 1048, 1096, 1152 and 1202 mm) on the productive performance of spineless cactus cultivated in semiarid environment and check that there were no differences in number of cladodes and in fresh and dry annual biomass between treatments, revealing that the increase of irrigation has not contributed to increase the yield of crop. Flores-Hernández et al. [68] also found that supplemental irrigation (740, 1060 and 1380 mm) did not provide increments cladodes production and dry matter productivity.

		Productivity (tons of forage/ha)	
Municipality	**Cutting period**	***Opuntia ficus indica* Mill**	***Nopalea cochenillifera* Salm Dyck**
Apodi	2 years	500	400
Cruzeta	2 years	215	200
Pedro Avelino	1 year	200	220

Source: Adapted from Ref. [66].

Table 3. Productivity of spineless cactus dense, fertilized and irrigated in Rio Grande do Norte State, Brazil.

Oliveira et al. [41] reported that regions with rainfall above 1000 mm/year can result in low productivity of spineless cactus, possibly due to excessive rainfall. Thus, the good yield of crops in semiarid is associated with fact that they need far less water than other conventional crops. In this case, spineless cactus produces well in areas with annual precipitation of up to 750 mm [14]. These results lead to the understanding that the productive benefits of increased of water blade for spineless cactus are more apparent in regions with very low rainfall levels compared to regions where the rainfall values exceed 750 mm [67, 68].

5.2.1. Management of spineless cactus in irrigated soil

Irrigation—For the choice of an irrigation system, some aspects should be considered, such as the quantity and quality of water, climate, topography, soil and crop to be irrigated; in other words, there is no irrigation system and yes, the one that more fits the conditions of resources available on the property [63]. In general, the most widely used irrigation system is located by drip, with a line per row [62, 67, 68]. The origin of water can be from various sources, from the dam to the wastewater or saltwater, where positive results were obtained on productivity [69]. Whatever the source of available water, to irrigate one hectare of spineless cactus, you will need a volume diary minimum of 5000 liters [63].

Planting area—The spineless cactus is a relatively demanding crop about physical-chemical characteristics of soil. Fertile soils, plants, and deep with sandy to clay texture should be selected, being most recommended clay-sandy soils. In the old days, it was common for the producers to choose the worst soils to plant the spineless cactus for being a very tough plant. However, for spineless cactus plantation irrigated the thought should be exactly the opposite, due to the high cost of the system, the high density of planting, and nutrient extraction, and to be a permanent crop, one must choose the best soil possible [63]. However, since provided that the soil is decompressed and organic matter is added, other types of soil can be used [2]. To do this, it must carry out an analysis of soil of area chosen, avoiding acid and salinized soil, choosing preferably light soils of gentle topography, and avoiding those shallow and stony. The analysis shall include both physical-chemical characteristics soil [63]. It is very important that the soil has good drainage, since very moist soils do not lend themselves to cultivation of spineless cactus [22].

Choice of cladodes—To select the cladodes in middle of the plants, avoid very small cladodes, young and very thin, as they have high mortality and low sprouting. Always cut the cladodes at the junction with sharp and clean knives. The custom of breaking cladodes manually forcing and twisting in joints should be avoided, because it impairs the healing of the cut and favors the installation of fungi. The cladodes should be inspected to ensure the absence of cochineal and rotting. The cladodes must undergo a wilt (the shade) average of 12 days for healing of cuts and loss of part of the water. To avoid contamination by fungi can be used any copper-based fungicide on the cut (20 g/20 L) or Bordeaux [63].

Planting spacing—Generally, it uses 1.4–2.0 m spacing between lines of planting and 10–30 cm, between the plants within line. A denser planting for *Nopalea cochenillifera* is recommended and less dense to plants of genus *Opuntia*. It is important to ensure a spacing of at least 20–25 cm between cladodes in line to allow clean with hoe. Examples of average densities would be spacing as 1.6 m × 0.25 m (25,000 plants/ha) or 1.4 m × 0.25 m (28,600 plants/ha). Higher densities can be found in 2.0 m × 0.10 cm spacing (50,000 plants/ha) or 1.4 m × 0.10 m (71,400 plants/ ha). An indication for systems in double rows irrigated is 1.80 × 0.50 × 0.40 m spacing to spineless cactus of genus *Opuntia* and 1.80 × 0.50 × 0.25 m for *Nopalea cochenillifera* [63].

Planting techniques—Traditionally, the spineless cactus planting in dry soil is performed 30–60 days before the rainy season. However, with the use of irrigation, the spineless cactus can be planted practically any time of year, since the cladodes are subjected to wilt and not be placed in wet soils. The organic and phosphorus fertilization must be deposited in bottom of groove, topped with a bit of land to avoid contact with the cladode. It is recommended that in clay soil the organic fertilizing is not placed at bottom of groove at planting, because it can provide the proliferation of fungi and cause rot of cladode. In this case, the organic fertilizing should be done later, spread among the ranks of planting during the rainy period, or when used in groove the manure should be cured. The position of planting of cladodes in groove or pit can be tilted (45°) or vertically, with the cut facing the soil. The form of planting most often used is burying1/3 of cladode. In the case of irrigated system, it is recommended to direct the wide face of cladode in east-west direction for it to make the most of the Sun's radiation to stimulate photosynthesis, sprouting and rooting [63].

Organic and chemical fertilization—The spineless cactus features a large response to organic fertilizing that must be applied in quantities of 20–40 tons/ha of cattle manure, goats or sheep, or 100 kg of manure for each ton of fresh matter produced. Thus, for a production of 300 tons FM/ ha would require 30 tons of manure. With the high productivity achieved by spineless cactus, the extraction of nutrients from soil is quite high and if these are not replenished, it may result in depletion of the soil. The five soil nutrients that appear to exert greater effect on performance of *Opuntias* are N, P, K, B and Na. How fertilizers have high cost, it is necessary to undertake a soil analysis for to know which nutrients that are disabled and apply them in the right quantities for each situation. When forward the soil analysis, one must ask the recommendation of fertilization to cultivation of spineless cactus [63].

Care with crops—Spineless cactus should be treated as crop, and since the producer will make a relatively high investment with the irrigated system deployment, every care should be taken to keep the terrain free of invasive plants. For that, at least three cleanings of terrain a year are required. If three cleanings cannot be held completely, at least one cleaning between lines should be made. Some herbicides have been used to facilitate the work, but so far there is no official indications of products to be used in the control of invasive plants in planting of spineless cactus [63].

Pest and disease control—The two major pests that affect spineless cactus are the scale cochineal and carmine cochineal. In this case can be used the same methods above of control for spineless cactus management in drought conditions [7, 12].

Cutting intensity in harvest—Traditionally, in Brazilian semiarid region, the spineless cactus is handled in drought conditions with the realization first cut to 2 years' age after planting and subsequent cuts every 2 years. With the use of irrigation and fertilization organic and chemistry, as well as low intensity management, can perform the first cut to 12 months and the subsequent cuts according to the need of forage. As most producers does not provide the ideal conditions of management and fertilizing soil, even with irrigation it would be wise to perform the first cut between 18 and 24 months, to consolidate the establishment of spineless cactus and then annual cuts. The results of research with the cultivars *Opuntia* and *Nopalea* proved that higher cuts, preserving even the secondary cladodes, produced 55% more than the cut while preserving the primary cladodes and 144% more when only the mother-cladode was left. This is another important management practice, because many producers practice very intense cuts, leaving only the mother-cladode. In more up cuts, even losing the part of production that is in field in first cut, the subsequent yields are highly compensators and the longevity and sustainability of spineless cactus is much favored [63].

6. Final considerations

The spineless cactus can achieve high productivity if handled correctly, with proper planting system, cultural practices, intensity and frequency which takes into consideration the photosynthetic capacity of culture, ensuring the animal supplementation.

Although be adapted to the edaphoclimatic conditions of the Brazilian Semiarid, the spineless cactus is demanding in cool night temperature and high relative humidity of the air for the good development.

In nonirrigated soil, the spineless cactus can present high productivity when compared to other traditional crops. However, in certain semiarid regions, often are necessary irrigation events in its production system so that it can meet its water needs for achieving high productivity.

Author details

Wilma Cristina Cavalcante dos Santos Sá[1]*, Edson Mauro Santos[1], Juliana Silva de Oliveira[1] and Alexandre Fernandes Perazzo[2]

*Address all correspondence to: wilmacristinasa@gmail.com

1 Federal University of Paraiba, Areia, PB, Brazil

2 State University of the Southwest of Bahia, Brazil

References

[1] Stintzing FC, Carle R. Cactus stems (Opuntia spp.): A review on their chemistry, technology, and uses. Molecular Nutrition & Food Research. 2005;**49**:175-194. DOI: 10.1002/mnfr.200400071

[2] Lopes EB, editor. Palma Forrageira: Cultivo, Uso Atual e Perspectivas de Utilização no Semiárido Nordestino. 1st ed. João Pessoa: EMEPA/FAEPA; 2007. 130 p

[3] Santos DC, Araújo LF, Lopes EB, Vasconcelos MF. Uso e aplicações da palma forrageira. In: Lopes EB, editors. Palma Forrageira: Cultivo, Uso Atual e Perspectivas de Utilização no Semiárido Nordestino. 1st ed. João Pessoa: EMEPA/FAEPA; 2007. pp. 57-89

[4] Simões DA, Santos DC, Dias FM. Introdução da palma forrageira no Brasil. In: Menezes RSC, Simões DA, Sampaio EVSB, editors. A palma no Nordeste do Brasil: conhecimento atual e novas perspectivas de uso. 1st ed. Recife: Editora Universitária da UFPE; 2005. p. 13

[5] Lira MA, Santos MVF, Dubeux Jr JCB. Meio século de pesquisa com a palma forrageira (Opuntia e Nopalea) - Ênfase em manejo. In: Zootec 2006; Recife. Pernambuco: Zootec; 2006

[6] Lima UC. Histórico e evolução da palma forrageira como recurso alimentar no nordeste brasileiro: uma revisão [thesis]. Paraíba: Universidade Estadual da Paraíba/Centro de Ciências Humanas e Agrárias; 2014

[7] Santos MVF, Lira MA, Dubeux Jr JCB, Ferreira MA, Cunha MV. Palma forrageira. In: Fonseca DM, Martuscello JA, editors. Plantas forrageiras. 1st ed. Viçosa: Editora UFV; 2010. p. 459

[8] Lima IMM, Gama NS. Registro de plantas hospedeiras (cactaceae) e de nova forma de disseminação de Diaspis echinocacti (Bouché) (Hemíptera: Diaspididae), cochonilha-da-palma-forrageira, nos estudos de Pernambuco e Alagoas. Neotropical Entomology. 2001;**30**:479-481. DOI: 10.1590/S1519-566X2001000300025

[9] Figueiredo VS, Silva EC, Gomes Filho MF. Sustentabilidade ambiental para o Semiárido paraibano: a busca de estratégias para o município de Juazeirinho - PB. In: XVI Encontro Nacional dos Geógrafos; 2010; Rio Grande do Sul: Associação dos Geógrafos Brasileiros; 2010. p. 11

[10] Dubeux Jr JCB, Araújo Filho JT, Santos MVF, Lira MA, Santos DC, Pessoa RAS. Adubação mineral no crescimento e composição mineral da palma forrageira - Clone IPA-201. Revista Brasileira de Ciências Agrárias. 2010;**5**:129-135. DOI: 10.5039/agraria.v5i1a591

[11] Oliveira EA, Junqueira SF, Mascarenhas RJ. Caracterização físico-química e nutricional do fruto da palma (Opuntia fícus indica L. Mill) cultivada no sertão do sub-médio São Francisco. Holos. 2011;**27**:113-119. DOI: 10.15628/holos.2011.517

[12] Santos MVF, Cunha MV, Lira MA, Dubeux Jr JCB, Freire JL, Pinto MSC, Santos DC, Souza TC, Silva MC. Manejo da palma forrageira. In: 2° Congresso Brasileiro de Palma e outras Cactáceas; 24-27 October 2011; Garanhuns. Pernambuco: Universidade Federal Rural de Pernambuco; 2011. pp. 1-15

[13] Farias I, Santos DC, Dubeux Jr JCB. Estabelecimento e manejo da palma forrageira. In: Menezes RSC, Simões DA, Sampaio EVSB, editors. A palma no Nordeste do Brasil: conhecimento atual e novas perspectivas de uso. 1st ed. Recife: Editora Universitária da UFPE; 2005. p. 81

[14] Silva RR, Sampaio EVSB. Palmas forrageiras Opuntia fícus-indica e Nopalea cochenillifera: sistemas de produção e usos. GEAMA. 2015;**2**:131-141

[15] Galvão Jr JGB, Silva JBA, Morais JHG, Lima RN. Palma forrageira na alimentação de ruminantes: cultivo e utilização. Acta Veterinaria Brasilica. 2014;**8**:78-85. DOI: 10.21708/avb.2014.8.2.3490

[16] Rocha JES. Palma forrageira no Nordeste do Brasil: Estado da arte. 1st ed. Sobral: Embrapa Caprino e Ovinos; 2012. 40 p

[17] Albuquerque SG. Cultivo da palma forrageira no Sertão do São Francisco [Internet]. 2000. Available from: https://www.embrapa.br/busca-de-publicacoes/-/publicacao/134098/cultivo-da-palma-forrageira-no-sertao-do-sao-francisco [Accessed: 11 May 2017]

[18] Silva CCF, Santos LC. Palma Forrageira (Opuntia Fícus-Indica Mill) como alternativa na alimentação de ruminantes. Revista Eletrônica de Veterinária. 2006;**7**:1-13

[19] Ferreira MA, Pessoa RAS, Silva FM, Bispo SV. Palma forrageira e uréia na alimentação de vacas leiteiras. 1st ed. Recife: EDUFRPE; 2011. 40 p

[20] Valadares Filho SC, Magalhães KA, Rocha Jr VR. Tabelas brasileiras de composição de alimentos para bovinos. 2nd ed. Viçosa: Universidade Federal de Viçosa, DZO; 2006. 329 p

[21] Ferreira MA. Palma forrageira na alimentação de bovinos leiteiros. 1st ed. Recife: Universidade Federal Rural de Pernambuco; 2005. 68 p

[22] Santos DC, Farias I, Lira MA, Santos MVF, Arruda GP, Coelho RSB, Dias FM, Melo JN. Manejo e utilização da palma forrageira (Opuntia e Nopalea) em Pernambuco. 1st ed. Recife: Empresa Pernambucana de Pesquisa Agropecuária - IPA; 2006. 33 p

[23] Souza CMS, Medeiros AN, Furtado DA, Batista AMV, Pimenta Filho EC, Silva DS. Desempenho de ovelhas nativas em confinamento recebendo palma-forrageira na dieta na região do semiárido nordestino. Revista Brasileira de Zootecnia. 2010;**39**:1146-1153. DOI: 10.1590/S1516-35982010000500028

[24] Pessoa RAS, Ferreira MA, Lima LE, Lira MA, Véras ASC, Silva AEVN, Sosav MY, Azevedo M, Miranda KF, Silva FM, Melo AAS, López ORM. Desempenho de vacas leiteiras submetidas a diferentes estratégias alimentares. Archivos de Zootecnia. 2004;**53**:309-320

[25] Araújo PRB, Ferreira MA, Brasil LHA, Santos DC, Lima RMB, Véras ASC, Santos MVF, Bispo SV, Azevedo M. Substituição do milho por palma forrageira em dietas completas para vacas em lactação. Revista Brasileira de Zootecnia. 2004;**33**:1850-1857

[26] Wanderley WL, Ferreira MA, Andrade DKB, Véras ASC, Farias I, Lima LE, Dias AMA. Palma forrageira (Opuntia ficus indica Mill) em substituição à silagem de sorgo (Sorghum bicolor (L.) Moench) na alimentação de vacas leiteiras. Revista Brasileira de Zootecnia. 2002;**31**:273-281. DOI: 10.1590/S1516-35982002000100030

[27] Mattos CW, Carvalho FFR, Guim A, Araújo GGL, Ribeiro VL, Araújo RFSS. Consumo de nutrientes de cordeiros Santa Inês alimentados com níveis crescentes de palma forrageira em dietas à base de feno de erva-sal. In: VI Congresso Nordestino de Produção Animal; 29 November 2010 - 02 December 2010; Mossoró. Rio Grande do Norte: Universidade Federal Rural do Semiárido; 2010. pp. 1-4

[28] Lima GFC, Silva JGM, Aguiar EM, Teles MM. Reservas forrageiras estratégicas para a pecuária familiar no semiárido: palma, fenos e silagem. 1st ed. Natal: EMPARN; 2010. 53 p

[29] Saag LMK, Sanderson GR, Moyna P, Ramos G. Cactaceae mucilage composition. Journal of the Science of Food and Agriculture. 1975;**26**:993-1000. DOI: 10.1002/jsfa.2740260716

[30] Sáenz C, Sepúlveda E, Matsuhiro B. Opuntia spp mucilage's: a functional component with industrial perspectives. Journal of Arid Environments. 2004;**57**:275-290. DOI: 10.1016/S0140-1963(03)00106-X

[31] Nogueira MS, Santos EM, Araújo GGL, Pinho RMA, Nunes CSSM, Parente HN. Ensilagem de palma forrageira. In: Santos EM, Parente HN, Oliveira JS, Parente MOM, editors. Ensilagem de plantas forrageiras para o Semiárido. 1st ed. São Luís: Editora EDUFMA; 2016. p. 249

[32] Nogueira MS. Perfil fermentativo e composição química de silagens de palma forrageira enriquecidas com fontes proteica, energética [thesis]. Paraíba: Universidade Federal da Paraíba; 2015

[33] McDonald P, Henderson AR, Heron SJE. The biochemistry of silage. 2nd ed. Marlow: Chalcombe Publicatins; 1991. 340 p

[34] Sá WCCS, Santos EM, Araújo GGL, Oliveira JS, Pinho RMA, Pereira GA, Brito GSMS, Silva EG. Perfil fermentativo de silagens de rações completas a base de palma forrageira. In: XXV Congresso da Associação Latino-Americana de Produção Animal/XI Congresso Nordestino de Produção Animal; 7-10 November 2016; Recife. Pernambuco: Universidade Federal Rural de Pernambuco; 2016. p. 1

[35] Brito GSMS, Sá WCCS, Oliveira JS, Bezerra HFC, Santos APM, Pereira GA, Corrêa YR, Pereira DM. Perfil fermentativo da silagem de palma forrageira com aditivos químico e microbiano. In: XXV Congresso da Associação Latino-Americana de Produção Animal/XI Congresso Nordestino de Produção Animal; 7-10 November 2016; Recife. Pernambuco: Universidade Federal Rural de Pernambuco; 2016. p. 1

[36] Arens K. Considerações sobre as causas do xeromorfismo foliar. Departamento de Botânica da Universidade de São Paulo. 1958;**15**:25-56

[37] Ferri MG. Evolução do conceito de xerofitismo. Departamento de Botânica da Universidade de São Paulo. 1960;**19**:103-113

[38] Lousã M, Monteiro A, Santos DE, Sousa E, Costa JC. Módulo de botânica: Manual de teóricas e práticas [Internet]. 2007. Available from: http://www.isa.utl.pt/files/pub/ensino/cdocente/MANUAL_BOTANICA_Fev2007.pdf [Accessed: 24 January 2017]

[39] Peixoto CP, Santos JMS, Poelking VGC. Curso de fisiologia vegetal [Internet]. 2013. Available from: https://www.passeidireto.com/arquivo/1907264/curso-de-fisiologia-vegetal [Accessed: 11 May 2017]

[40] Silva TGF, Primo JTA, Silva SMS, Moura MSB, Santos DC, Silva MC, Araújo JEM. Indicadores de eficiência do uso da água e de nutrientes de clones de palma forrageira em condições de sequeiro no Semiárido brasileiro. Bragantia. 2014;**73**:184-191. DOI: 10.1590/brag.2014.017

[41] Oliveira FT, Souto JS, Silva RP, Andrade Filho FC, Pereira Jr EB. Palma forrageira: adaptação e importância para os ecossistemas áridos e semiáridos. Revista Verde. 2010;**5**:27-37

[42] Hills FS. Anatomia e morfologia. In: Barbera G, Inglese P, editors. Agroecologia, cultivos e usos da palma forrageira. 2nd ed. João Pessoa: SEBRAE/PB; 2001. pp. 28-34

[43] Taiz L, Zeiger E. Fisiologia Vegetal. Translation: Santarém ER, Mariath JEA, Astarita LV, Dillenburg LR, Rosa LMG, Oliveira PL. 3rd ed. Porto Alegre: Artmed; 2004. 719 p

[44] Nobel PS. Biologia ambiental. In: Barbera G, Inglese P, editors. Agroecologia, cultivos e usos da palma forrageira. 2nd ed. João Pessoa: SEBRAE/PB; 2001. pp. 36-48

[45] Pereira AR, Angelocci LR, Sentelhas PC. Agrometeorologia: fundamentos e aplicações [Internet]. 2002. Available from: https://www.passeidireto.com/arquivo/3817545/livro-agrometeorologia-pereira-angelocci-sentelhas [Accessed: 24 January 2017]

[46] Souza LSB, Moura MSB, Silva TGF, Soares JM, Carmo JFA, Brandão EO. Indicadores climáticos para o zoneamento agrícola da palma forrageira (Opuntia sp.) [Internet]. 2008. Available from: http://www.alice.cnptia.embrapa.br/handle/doc/156105 [Accessed: 24 January 2017]

[47] Moura MSB, Souza LSB, Silva TGF, Sá IIS. Zoneamento agroclimático da palma forrageira para o Estado de Pernambuco. 1st ed. Petrolina: Embrapa Semiárido; 2011. 28 p

[48] Ministério da Agricultura, Pecuária e Abastecimento - MAPA. Portaria n° 295, de 3 de agosto de 2011 [Internet]. Available from: http://sistemasweb.agricultura.gov.br/sislegis/action/detalhaAto.do?method=visualizarAtoPortalMapa&chave=485706458 [Accessed: 11 May 2017]

[49] Bezerra BG, Araújo JS, Pereira DD, Laurentino GQ, Silva LL. Zoneamento agroclimático da palma forrageira (Opuntia sp.) para o estado da Paraíba. Revista Brasileira de Engenharia Agrícola e Ambiental. 2014;**18**:755-761. DOI: 10.1590/S1415-43662014000700013

[50] Inglese P. Plantação e manejo do pomar. In: Barbera G, Inglese P, editors. Agroecologia, cultivos e usos da palma forrageira. 2nd ed. João Pessoa: SEBRAE/PB; 2001. pp. 79-92

[51] Silva TGF, Araújo Primo JT, Morais JEF, Diniz WJS, Souza CAA, Silva MC. Crescimento e produtividade de clones de palma forrageira no semiárido e relações com variáveis meteorológicas. Revista Caatinga. 2015;**28**:10-18

[52] Almeida J, Peixoto CP, Ledo CAS. Desempenho vegetativo e produtivo da palma forrageira. Enciclopédia Biosfera. 2012;**8**:571-581

[53] Silva JA, Bonomo P, Donato SLR, Pires AJV, Rosa RCC, Donato, PER. Composição mineral em cladódios de palma forrageira sob diferentes espaçamentos e adubações química. Revista Brasileira de Ciências Agrárias. 2012;**7**:866-875. DOI: 10.5039/agraria.v7isa2134

[54] Dubeux Jr JCB, Santos MVF. Exigências nutricionais da palma forrageira. In: Menezes RSC, Simões DA, Sampaio EVSB, editors. A palma no Nordeste do Brasil: conhecimento atual e novas perspectivas de uso. 1st ed. Recife: Editora Universitária da UFPE; 2005. p. 105

[55] Dubeux Jr JCB, Santos MVF, Lira MA, Santos DC, Farias I, Lima LE, Ferreira RLC. Productivity of Opuntia fícus-indica (L.) Miller under different N and P fertilization and plant population in north-east Brazil. Journal of Arid Environments. 2006;**67**:357-372. DOI: 10.1016/j.jaridenv.2006.02.015

[56] Ramos JPF, Leite MLMV, Oliveira Jr S, Nascimento JP, Santos EM. Crescimento vegetativo de Opuntia fícus-indica em diferentes espaçamentos de plantio. Revista Caatinga. 2011;**24**:41-48

[57] Guimarães AS, Leite JRA, Sales ECJ, Alves DD, Albuquerque CJB. Implantação de lavouras de palma forrageira [Internet]. 2014. Available from: www.epamig.br/index.php?option=com_docman&task=doc_download&gid [Accessed: 11 May 2017]

[58] Menezes RSC, Simões DA, Sampaio EVSB, editors. A palma no Nordeste do Brasil: conhecimento atual e novas perspectivas de uso. 1st ed. Recife: Editora Universitária da UFPE; 2005. 258 p

[59] Farias I, Lira MA, Santos DC, Tavares Filho JJ, Santos MVF, Fernandes APM, Santos VF. Manejo de colheita e espaçamento da palma forrageira, em consórcio com sorgo granífero, no Agreste de Pernambuco. Pesquisa Agropecuária Brasileira. 2000;**35**:341-347. DOI: 10.1590/S0100-204X2000000200013

[60] Teles MM, Santos MVF, Dubeux Jr JCB, Bezerra Neto E, Ferreira RLC, Lucena JEC, Lira MA. Efeitos da Adubação e de Nematicida no Crescimento e na Produção da Palma Forrageira (Opuntia ficus indica Mill) cv. Gigante. Revista Brasileira de Zootecnia. 2002;**31**:52-60. DOI: 10.1590/S1516-35982002000100006

[61] Silva NGM. Produtividade, morfometria e acúmulo de nutrientes da palma forrageira sob doses de adubação orgânica e densidades de plantio [thesis]. Recife: Universidade Federal Rural de Pernambuco, Departamento de Zootecnia; 2012

[62] Pereira PC. Evapotranspiração e crescimento de clones de palma forrageira irrigada no semiárido brasileiro [thesis]. Viçosa: Universidade Federal de Viçosa, Departamento de Engenharia Agrícola; 2013

[63] Lima GFC, Wanderley AM, Guedes FX, Rego MMT, Dantas FDG, Silva JGM, Novaes LP, Aguiar EM. Palma Forrageira Irrigada e Adensada: uma Reserva Forrageira Estratégica para o Semiárido Potiguar. 1st ed. Parnamirim: EMPARN; 2015. 62 p

[64] Castillo IO, Cueto WJA, Santamaría JC, Murillo BA, Troyo ED. N, P and K Removal by Vegetable Cactus Pear Cultivated with Dairy Manure under Drip Irrigation. Acta Horticulturae. 2006;**728**:193-198

[65] Ferreira PA, Garcia GO, Miranda GV, Oliveira FG, Santos DB. Tolerância da variedade de milho ufvm 100 à salinidade avaliada por três métodos. Irriga. 2007;**12**:532-544

[66] Empresa de Pesquisa Agropecuária do Rio Grande do Norte - EMPARN. Palma forrageira adensada e irrigada [Internet]. Available from: http://adcon.rn.gov.br/ACERVO/EMPARN/DOC/DOC000000000018672.PDF [Accessed: 20 January 2017]

[67] Queiroz MG, Silva TGF, Zolnier S, Silva SMS, Lima LR, Alves JO. Características morfofisiológicas e produtividade da palma forrageira em diferentes lâminas de irrigação. Revista Brasileira de Engenharia Agrícola e Ambiental. 2015;**19**:931-938. DOI: 10.1590/1807-1929/agriambi.v19n10p931-938

[68] Flores-Hernández A, Castillo IO, Amador BM. Yield and physiological traits of prickly pear cactus 'nopal' (Opuntia spp.) cultivars under drip irrigation. Agricultural Water Management. 2004;**70**:97-107. DOI: 10.1016/j.agwat.2004.06.002

[69] Dantas FDG. Lâminas de água salina e doses de adubação orgânica na produção de palma miúda adensada no Semiárido [thesis]. Macaíba: Universidade Federal do Rio Grande do Norte, Unidade Acadêmica Especializada em Ciências Agrárias; 2015

Chapter 3

Development of Tropical Forages in Veracruz, Mexico: Agronomic Approach for the New Forage Legume *Cratylia argentea*

Braulio Valles-de la Mora,
Epigmenio Castillo-Gallegos, Jesús Jarillo-Rodríguez,
Eliazar Ocaña-Zavaleta and
Miguel Angel Alonso-Díaz

Additional information is available at the end of the chapter

http://dx.doi.org/10.5772/intechopen.69617

Abstract

Pastures in the coast of the Gulf of Mexico are characterized by native species (*Paspalum* spp., *Axonopus* spp., etc.). These, are limited by productivity due to their low nutritional quality and poor persistence on grazing. The adaptation of new grass and legume species is essential to improve the productivity of animal production. An initial assessment should include the climatic and edaphic adaptation to the region. Species, such as *Andropogon gayanus, Pueraria phaseoloides, Centrosema* spp., *Arachis pintoi,* and Cratylia argentea, were evaluated, showing encouraging results compared to native species; our efforts were focused on *C. argentea*. Several research methods were applied to meet the objectives outlined for each experiment, including methodologies for the establishment of new species. All these trials were subject to rigorous experimental designs, and data were analyzed statistically, using the most adequate programs. These experiences allow us to visualize the most promising materials for the specific conditions of climate and soil. The potential results of this new forage species stand out. Also, these experiments allowed the development of new management practices to improve the productivity of the animal production systems of the region. *C. argentea* demonstrated its high forage value as a species suitable for silvopastoral systems.

Keywords: tropical pastures, edaphic and climatic adaptation, Cratylia argentea, Veracruz, México

1. Introduction

Livestock in the tropical region of Mexico has pastures composed of *Paspalum* spp., *Axonopus* spp., and legumes of the genus *Desmodium, Indigofera, Centrosema,* and *Mimosa,* called "native grams" [1, 2]. These species appear in natural form after cut-down and burning of the original forest [3]. However, the animal production achieved with these species is 50% lower than that obtained with introduced grasses. Daily growth rates for these species do not regularly exceed 25 kg DM/ha/day, in addition to showing a marked seasonal growth, which limits sustainable levels of dairy and meat production. This seasonality is the result of variations in climate during the year, especially in the winter or "Nortes" (November to February). This season is critical for forage production due to the low temperatures for tropical species (around 16°C), high cloudiness, and monthly precipitation of 100 mm [2]. Also, overgrazing contributes to nutrient and soil organic matter losses, partly because farmers do not fertilize their pastures due to the high costs of fertilizers and due to the low response to this practice, in terms of forage yield.

Due to this problem, it is necessary to find forages that adapt to these critical times. There are some legume species adapted to the conditions of the dry season, which have already been tested in other tropical regions of Latin America [4], so it is possible that some of them could be established in the central region in the state of Veracruz (Mexico).

Tropical forage legumes could be alternative solutions to these problems, since they present high nutritional quality and can also fix N to the soil, which, over time, becomes available to the associated grasses, increasing the production of pastures. Moreover, legumes in association with grasses can increase the amount of charcoal sequestered by pastures [5, 6]. Therefore, these plants can contribute to diminish the negative impact that the pastoral industries have on the environment.

Species, such as *Cratylia argentea* could be evaluated under grazing, associated to native or introduced grasses, or as a protein bank, however, their capacity to improve pasture production and productivity in this region could be verified [7]. *C. argentea* is a shrub legume native to Brazil, Peru, and Bolivia, which adapts well from sea level up to 900 m asl, in places with humid or subhumid climates, and dry periods of five to six months. It is also adapted to acidic soils of medium fertility with good drainage [8]. The accession CIAT 18516 is the most evaluated and can be harvested every 12–14 weeks yielding from 8 g of MS/plant in the municipality of Isla, Veracruz. It grows well in the dry season, producing about 30–50% of annual forage yield.

2. First evaluations of *C. argentea* under cutting regime

2.1. Total dry matter and nutritive quality of four *C. argentea* accessions after a year of the establishment period

2.1.1. Reasoning

C. argentea shows an abundant growth during its establishment period that usually lasts one year during which high-quality forage can be harvested to be used (fresh or dry) as a grazing herd supplement, or included in dry rations. In addition, numerous nonedible stems accumulate which can be dried and used as fuelwood in rural homes.[1]

For this reason, it is speculated that it is necessary to evaluate—in a first stage—the production of edible fodder as well as the secondary aspects, such as inedible biomass production, which could be an energy source.

2.1.2. Materials and methods

The objective of this experiment was to evaluate—under conditions of warm and humid climate and acid soils—the forage yield and the nutritive quality of the accessions of *C. argentea* CIAT 18516, 18666, 18668, and 18676 to the cut of establishment, after almost a year of uninterrupted growth.

The study was carried out at the F1 Heifer Production Unit of the Center for Teaching, Research and Extension in Tropical Livestock (CEIEGT, its acronym in Spanish), Faculty of Veterinary Medicine and Animal Husbandry, National Autonomous University of Mexico. The unit is located in the municipality of Atzalan, state of Veracruz (Mexico), at 20°02' N latitude, 97°34' W length and 111 m above sea level.

The minimum and maximum temperature averages, in addition to rain, during the experimental period were 20.0, 30.4°C, and 1.926 mm, respectively. The soil texture in the first 20 cm is 52, 28, and 20% clay, silt, and sand, respectively, with an acidity of pH 4.7 and with low total N (0.983 $g \cdot kg^{-1}$), extractable P (0.04 $cmol \cdot kg^{-1}$), extractable K (1.45 $cmol \cdot kg^{-1}$), and cation exchange capacity (11.95 $cmol \cdot kg^{-1}$).

The experimental area was cultivated in a conventional manner, and sowing, using seed, was performed on September 1, 2006. The spacing between furrows and sowing sites was 1 m each. The first harvest of forage was made from August 23–27, 2007 at a cutting height of 70 cm. The experimental design was a randomized complete blocks, using the slope as criteria to block, with three blocks as repetitions.

The plant parts analyzed were leaves (L), edible stems (ES, <3 mm diameter), and nonedible stems (NES, >3 mm diameter). Plants were dried at 60°C, and milled at 2 mm. Samples used for gas in vitro methodology were milled at 1 mm.

Chemical analyses were performed for crude protein (CP, %); Kjeldahl [9]. The methodology of Van Soest et al. [10] was used for determining neutral detergent fiber (NDF, %), acid detergent fiber (ADF, %), and lignin (LIG, %).

The in situ disappearance (ISD, %) of leaves at 3, 6, 9, 12, 24, 48, and 72 h of ruminal incubation was done in triplicate in three rumen-fistulated cows using the nylon bag technique [11], without pretreatment with pepsin in acid medium. The time-out disappearance was estimated in duplicate, washing with water at 38°C for 30 min.

[1]Data from this experiment were already published by: Castillo-Gallegos E., Estrada -Flores J.G., Valles-de la Mora B., Castellan-Ortega O.A., Ocaña-Zavaleta E., y Jarillo-Rodríguez J. 2013. Rendimiento total de materia seca y calidad nutritiva de hojas y tallos jóvenes de cuatro accesiones de *Cratylia argentea* en el trópico húmedo de Veracruz, México. Avances en Investigación Agropecuaria (México), 17(1):79-93. ISSN:0188789-0.

The data were adjusted to the model proposed by these same authors: y = a + b (1 − e (−c * t)), where "y" is the dry matter (%) degraded at time t, "a" is the highly soluble dry matter when t = 0, "b" is the slowly degradable (%) dry matter, "a + b" is the extent of digestion (%), "c" is the fractional rate of degradation of "b" (fraction/h), and "t" is the incubation time in the rumen (h).

The kinetics of in vitro gas production of edible leaves and stems was evaluated [12]. The generated data were adjusted to the exponential equation of Krishnamoorthy et al. [14]: y = b (1−e−c * (x−L)), where "y" (ml) is the accumulated gas production at time "x" (h), "B" is the asymptote or potential gas production accumulated as "x" → J (ml), "C" is the fractional rate at which gas production accumulates at time "x", and "L" is the lag time (the time h, that takes the ruminal microbes to colonize and initiate gas production from the NDF of slow degradability).

In order to meet the assumptions of the analysis of variance, the percentage units were transformed to "arcsin √%/100" and the consumable material ratio (L + ES) 84 to nonconsumable stems (NES), which is dimensionless, was transformed to values of "natural log of y + 1". The model of analysis of variance had the effects of block as a repetition (the cow confused with the block in the case of ISD), accession, component of the plant, and the interaction accession × plant component. Proc GLM of SAS was used to perform analyses. LS means option was used to generate minimum squares means and comparisons between them [13].

2.1.3. Results

The accession had no significant effect on the dry matter yield at the first cut of the L, ES, and NES components, which showed mean ± standard errors of 2580 ± 212, 33 ± 5, and 2444 ± 233 kg/ha, respectively. In contrast, the proportions of nonedible leaves and stems were significantly affected by the accession (**Table 1**). CIAT 18668 showed the highest HO ratio, which was not statistically different from CIAT 18516 and 18676, but statistically superior to CIAT 18666, which was the lowest of all accessions.

From the effects of the model, only the component of the plant was significant ($P < 0.05$) on all chemical components, whereas the other effects (block, accession, and the interaction ×

CIAT accession	Leaf	Edible stem	Nonedible stem	Edible/onedible
		%		
18516	48.96 ab	0.78 a	50.25 ab	0.99 ab
18666	45.92 b	0.66 a	53.40 a	0.88 b
18668	56.35 a	0.59 a	43.03 b	1.32 a
18676	54.65 ab	0.45 a	44.89 ab	1.23 ab

Means followed by the same letter are statistically similar ($P > 0.05$).

Table 1. Percentage of total yield of dry matter at first cut, occupied by leaf, edible stem, nonedible stem, and edible/nonedible ratio, of four accessions of *Cratylia argentea* grown in the humid tropics of the state of Veracruz, Mexico.

component of the plant) were not significant. The general means ± standard errors were: 19.10% ± 0.70% for CP, 61.10% ± 1.00% for NDF, 42.20% ± 1.20% for ADF, and 14.20% ± 0.30% for LIG. The leaves had significantly more CP than the edible stems (20% vs. 16.20%), and were significantly lower than stems in NDF (57.10% vs. 65.20%) and ADF (36.8% vs. 47.50%), but not in LIG, as the leaves showed a significantly higher content than the edible stems (15.10% vs. 13.30%).

With respect to the ISD of the leaf component, the effect of the block was not significant on "a" and "b", but it was on the "c" fractional rate. The effect of the accession was significant only on "a", but not on the other parameters. Likewise, neither the effect of the cow or the accession × cow interaction affected the parameters.

The average coefficient of determination of the individual curves was 0.8970 with a standard error of ± 0.0222. Therefore, the accessions only differed in the proportion of the highly soluble component of the dry matter: 29.97, 30.06, 33.15, and 31.32% for CIAT 18516, 18666, 18668, and 18676, respectively; 18668 being significantly higher than the others, which did not differ from each other; while all had a common fractional degradation rate (0.0488 ± 0.0192 per hour) of the slowly degradable dry matter component (30.60% ± 4.52%), as shown in **Figure 1**.

The parameters of this model [14] were not affected by the block, accession, or the interaction accession × component of the plant. The fit of the individual curves was quite good, given that the average of the determination coefficients was 0.9907 with a standard error of ± 0.0070. Therefore, a single curve could be used to describe the dynamics of in vitro gas production of the four accessions, which is presented in **Figure 2**, where it is also shown that the effect of the plant component was significant on all the parameters.

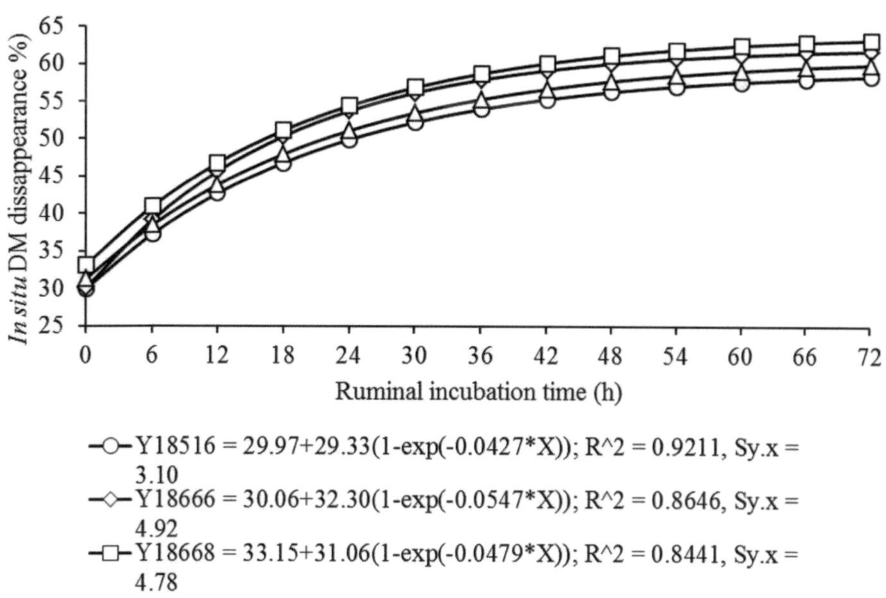

Figure 1. In situ dry matter disappearance as a function of the incubation time in the rumen, according to Ref. [11], of leaves of the first harvest of four accessions of *Cratylia argentea* cultivated in the humid tropics of Veracruz, Mexico.

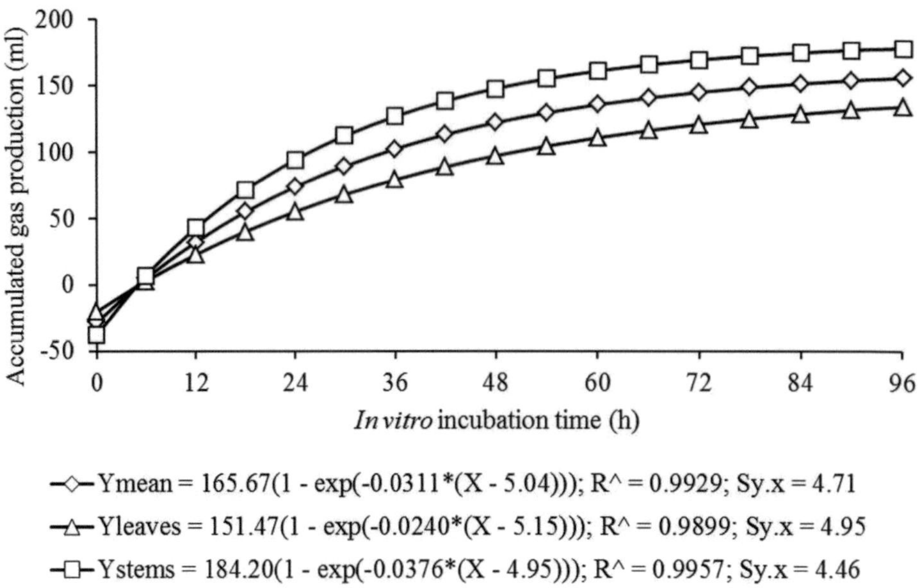

Figure 2. In vitro gas production as a function of incubation time [14] of leaves, edible stems, and both components, averaged through the four accessions of *Cratylia argentea* grown in the humid tropics of Veracruz, Mexico.

2.1.4. Discussion

The average dry matter yield of leaves, edible stems, and nonedible stems were statistically the same. The proportion of leaves was higher in CIAT 18668 than in CIAT 18666, which showed the lowest leaves ratio. The average dry matter yield of leaves, edible stems, and nonedible stems were statistically the same. On the other hand, CIAT 18516 and CIAT 18676 were similar to CIAT 18668, implying that the first two would be as good candidates to be selected as the latter (**Table 1**). In summary, the dry matter yield variables and the derived variables were only useful in selecting the least productive accession.

After nearly 12 months of uninterrupted growth, 51.50% of the aerial biomass were leaves, which resulted in a ratio of 1.1:1 with respect to edible material:nonedible stems. In Costa Rica, *C. argentea* was harvested every 12 weeks for a year, and the leaf:stem ratio was found to be 1.76:1 for "meson" soils and 1.43:1 for "terrace" soils [15]. The Costa Rican values are higher than those of the present experiment, because there was a difference of 40 weeks (52 vs. 12) in cutting age. In any case, ratios > 1:1 reflect the *C. argentea*'s ability to retain both young and mature leaves.

In the present study, the chemical composition of *C. argentea* foliage (**Table 2**) was fairly uniform despite changes in the environment and the management. This is a desirable feature to improve livestock production in low-quality pastures. The chemical composition of the 4 accessions did not change drastically with plant age and remained at acceptable levels after nearly 12 months of uninterrupted growth.

The accessions were statistically similar with respect to their contents of crude protein, neutral detergent fiber, acid detergent fiber, and lignin. The leaves in this aspect exceeded the

Botanical component	Chemical variables, %			
	CP	NDF	ADF	LIG
Leaves	20.0	57.1	36.8	15.1
Edible stems	16.2	65.2	47.5	13.3

CP= crude protein, NDF= neutral detergent fiber, ADF=acid detergent fiber, LIG=lignin.

Table 2. Chemical components of *C. argentea* leaves and edible stems after 364 days of uninterrupted growth, in the humid tropic of Veracruz, Mexico.

edible stems. In fact, since the contribution of the edible stems to the total dry matter yield was so small, only leaf yield should be considered to select the best accession.

Regarding the in situ dry matter degradability, of leaves and young stems, no statistical effect of the age of harvest on this variable was found. An experiment [16] reported that the parameter values model [11] for 2, 3, and 4 months of age were, respectively: for "a", 31.30, 28.20, and 24%; for "b", 30.30, 24.40, and 26.50%; for "c", 0.08, 0.08, and 0.07 per hour. The values of "a" and "b" are very similar to those of the present study (**Figure 1**), whereas those of "c" are higher by about 0.03 units; the difference may have been due to the higher proportion of mature leaves in the present experiment, in which the plants were harvested at an advanced stage of maturity.

The gas production dynamics were different between leaves and edible stems, the former having a lower gas production potential and fractional rate than the latter. Leaf tannins are known to interfere with the amount and rate of gas production [17]. Therefore, the leaves are digested at a slower rate and to a lesser extent than the young stems.

C. argentea has been classified as nontaniniferous [18]. These authors found that the nontannic legumes *Vigna unguiculata* and *C. argentea*, with 0% condensed tannins, had asymptotic gas production values (214 and 157 ml, respectively) and dry matter degradation (68.50 and 51.20%, respectively) higher, compared to hays of the taniniferous species *Calliandra calothyrsus* (23% of condensed tannins) and *Flemingia macrophylla* (3.66% of condensed tannins) which produced less gas (93 and 80 ml, respectively), and whose dry matter was less degraded (25.60 and 25%, respectively). Given these results, tannin determinations are unnecessary in *C. argentea*.

On the other hand, the two bioassays made in the edible material did not show any practical or statistical differences between accessions in terms of the in situ degradation of the dry matter or the in vitro gas production dynamics (**Figures 1, 2**).

2.1.5. Conclusions

None of the accessions was superior to the others. The four accessions of *C. argentea* were similar in yield of DM leaf, edible stem, and nonedible stems, as well as in the nutritive value of its edible constituents, which in leaf were close to 20% crude protein, which can be a good protein supplement, given fresh or dry, for animals grazing tropical grasses of low nutritional quality. New studies are needed, with cuts at different ages of regrowth and at different climatic seasons of the year, to identify the most productive accession.

2.2. Performance of *C. argentea* during three climatic seasons and several ages of cutting[2]

2.2.1. Reasoning

The use of shrub legumes with high nutritional quality that can thrive at the time of year when most grasses do not becomes an alternative to address the shortage of food in the dry period. However, not all the forage trees or shrubs yield enough amounts of biomass to feed cattle. The age of regrowth and climatic seasons are documented to affect the yield and forage quality of woody forage species [19].

Cutting forage trees at different seasons of the year (dry season vs. wet season) and at different stages of development (flowering vs. vegetative) may also influence subsequent regrowth. Many studies have reported that the highest total biomass yield is obtained in the longer harvest intervals. Accessions CIAT 18674 and CIAT 22406 were identified as promising for DM production, particularly in the dry season. In Quintana Roo, Mexico, an experiment [20] was carried out, evaluating several legumes. Among them *C. argentea* showed an effect of the season and cutting age on the dry matter yield of this legume. These authors observed that performance among species varied within each season. In the tropics of Mexico, native pastures are the basis of grazing for cattle. This type of vegetation is of low quality, and due to the climatic variations, it presents a high seasonality of its growth. This occurs regularly in the dry and wintry seasons. This situation has received very little attention in the Mexican humid tropics, so it needs to be evaluated. The objective of this study was to evaluate the effect of different regrowth ages on the yield and forage quality of four accessions of *C. argentea* in three climatic seasons.

2.2.2. Materials and methods

This experiment was carried out at the same site as that described previously, and the climatic conditions are shown in **Figure 3**.

On September 1, 2006, four forage accessions of *C. argentea* were established in 10 × 3 m plots, with an arrangement within them of 1 m distance between rows and within rows. The plots were subdivided into 4 areas, each corresponding to each cut age (6, 9, 12, and 15 weeks). Eleven months after this planting, a first cut was made to standardize the treatments. Later, the cuts corresponding to each age were made. The cut-off dates for each of the seasons were as follows: rainy season (October 10 and 29, November 24, and December 10, 2007); winter season (January 31, February 20, March 12, and April 2, 2008); and dry season (May 19, June 4 and 25, and July 15, 2008). The height of cutting (above ground level) was 70 cm for all cases. The following variables were evaluated: dry matter yield (DMY, $kg \cdot ha^{-1}$), crude protein (CP, $g \cdot kg^{-1}$ DM), neutral detergent fiber (NDF, %), acid detergent fiber (ADF, %), lignin (LIG, %), and 72 h in situ dry matter degradation (ISDMD, %).

[2]Data presented here are taken from: Valles-De la Mora, B., Castillo-Gallegos, E., Ocaña-Zavaleta, E., & Jarillo-Rodríguez., J.2014. *Cratylia argentea*: A potential fodder shrub in silvopastoral systems. Yield and quality of accessions according to regrowth ages and climatic seasons, *Revista Chapingo Serie Ciencias Forestales y del Ambiente*, XX(2) 277-293. http://dx.doi.org/10.5154/r.rchscfa.2013.11.040.

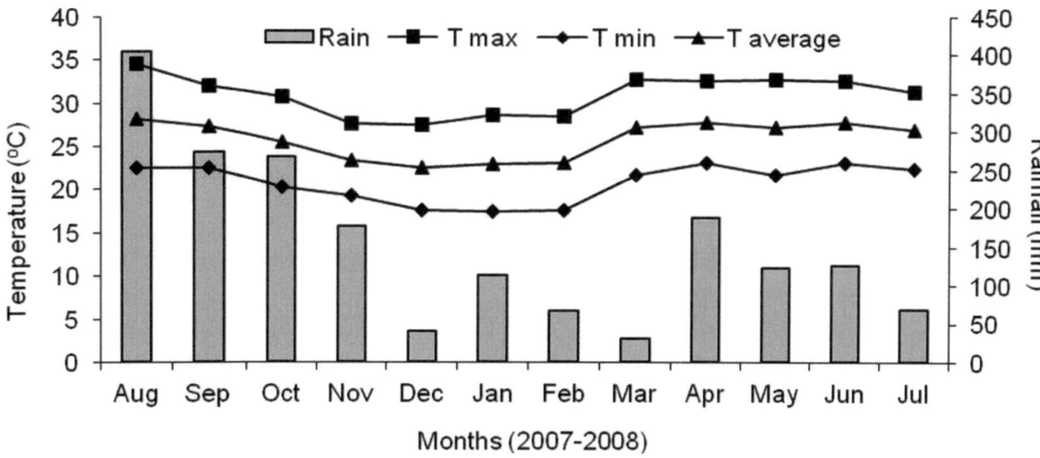

Figure 3. Maximum (T max), minimum (T min), and average temperature (T average), and rainfall (rain) during the experimental period in Veracruz, Mexico.

To analyze the harvested material in the laboratory, leaves (leaflets and petiole) and stems up to 3 mm were separated. These parts of the plant are considered the consumable material by cattle. Samples of these materials were analyzed to determine the percent dry matter (DM), crude protein [9], neutral detergent fiber (NDF), acid detergent fiber (ADF), and lignin (LIG) [10]. Also, DM yield was calculated [21]. In situ dry matter digestion was estimated, with incubation times of 3, 6, 9, 12, 24, 48, and 72 h [11]; degradation parameters were obtained by fitting the data to a model where, y = a + b (1 − e − ct), where y = DM degraded at time "t" (%), a = rapidly degradable fraction (intercept) (%), b = slowly degradable fraction (%), a + b = potentially degradable DM (extent of degradation) (%), c = rate of degradation (degradable fraction per hour), t = time of incubation in the rumen (h), and e = base of the natural logarithms [22].

A randomized complete block design and three replications (blocks) were applied as the experimental design for this experiment. We used the slope of the terrain as a criterion for blocking. Also, we assigned four plots (one per accession) to each one of the blocks. PROC MIXED of Statistical Analysis System [23] was performed for the ANOVA. The exponential growth model: $y = ae^{bx}$, where y = DMY ($kg{\cdot}ha^{-1}$), a = DMY when x = 0, b = rate constant expressed in inverse x units ($1{\cdot}x^{-1}$), x = age of regrowth in weeks, was used to adjust data. Also, for each season, a fitting process was done. Dry matter yield and quality variables were analyzed with PROC MIXED; and ANOVA with least squares means "t" test comparisons was performed. For ISDMD, a particular curve for each combination of accession and regrowth age was fit, so each parameter could be analyzed individually as the response variable in the analysis of variance.

2.2.3. Results and discussion

2.2.3.1. Estimations of dry matter yield for average age of regrowth corresponding to each season

For DMY, the analysis of variance resulted with statistical differences ($P < 0.0001$) corresponding to the effect of age of regrowth and season. Each one of the seasons was different among them: the means and standard error (±) for rainy, winter, and dry seasons were 2615 ± 188, 1783 ± 61, and 3632 ± 306 kg·ha^{-1}, respectively. These values represented an annual forage distribution of 33, 22, and 45%, respectively. Over the three seasons, the mean values per 18516, 18666, 18668, and 18676 accessions were 2311 ± 261, 3048 ± 321, 2567 ± 280, and 2781 ± 301 kg·ha^{-1}, respectively. The averages considering the three seasons at 6, 9, 12, and 15 weeks were: 1225, 29138, 3366, and 4062 kg·ha^{-1}, respectively. Regardless of season, forage production showed no statistical differences within regrowth ages. Also, the accessions behaved similarly (data not shown) (**Table 3**).

In the dry season, *C. argentea* and other legumes were evaluated [2] and were found to produce low yields averaging 0.6 t·ha ^{1}. This value is considerably lower than the 3.6 ± 0.32 t·ha^{-1} averaged over the four regrowth ages. Also, in the dry season in Venezuela, the same 4 accessions showed a range of forage production (leaves) from 651 to 862 kg·ha^{-1} per cut [24]. Other authors [7] in Isla, Veracruz, Mexico (summer rainfall, 1000 mm·year^{-1}) mentioned that in the dry season the total annual yield was only 25%, compared to 55% and 20% during the rainy and winter seasons, respectively. Also, in Anzoategui, Venezuela (rainfall, 1044 mm·year^{-1}), an experiment reported that these same accessions produced more forage in the rainy season, while the dry season's yield was only 37% of that achieved in the rainy season [25].

2.2.3.2. Crude protein content by season and age of regrowth

Levels of crude protein in *C. argentea* by season are shown in **Table 4**. Values of crude protein by season were: 224 ± 2.5 g·kg^{-1} DM, 263 ± 2.4 g·kg^{-1} DM, and 259 ± 5.8 g·kg^{-1} DM for the rainy, winter, and dry seasons, respectively. Linear regression equations (Y = a + bx) were developed for each accession in order to look for variations in this parameter, yielding the following results for the accessions 18516, 18666, 18668, and 18676: $Y = 24.17–0.007x$, $R^2 = 4.3668 \times 10^{-6}$; $Y = 22.69 + 0.198$, $R^2 = 0.54$; $Y = 22.31 + 0.27$, $R^2 = 0.75$; $Y = 22.82 + 0.25x$, $R^2 = 0.64$, respectively.

The content of crude protein shown here are different or similar to those found by other researchers. In Colombia (Antioquia) researchers reported that during dry season, the height

Season	18516	18666	18668	18676
	Average 6–15 weeks			
Rainy	2289 ± 374	3011 ± 358	2608 ± 278	2552 ± 486
	Exponential growth model: $Y = 842e^{0.1026}$, $R^{2*} = 0.50$, RSE = 937, n = 48			
Winter	1396 ± 245	2106 ± 411	1495 ± 298	1930 ± 314
	Exponential growth model: $Y = 440e^{0.1256}$, $R^2 = 0.45$, RSE = 840, n = 46			
Dry	3248 ± 544	4026 ± 721	3544 ± 620	3711 ± 610
	Exponential growth model: $Y = 873e^{0.1278}$, $R^2 = 0.56$, RSE = 1415, n = 48			

*R^2: coefficient of determination; RSE: residual standard error. ± Average standard error.

Table 3. Dry matter yield (DMY, kg ha^{-1}) of four *Cratylia argentea* accessions as average of four regrowth ages, in three climatic seasons.

		Cratylia argentea CIAT accessions			
Season	**Cutting age (weeks)**	**18516**	**18666**	**18668**	**18676**
			CP, %		
Rainy	6	25.1±0.5 [a]	24.5±0.5 [a]	23.5±0.3 [a]	25.0±0.8 [a]
	9	22.6±0.9 [ab]	22.9±0.5 [a]	23.2±0.6 [a]	22.9±0.8 [a]
	12	20.7±0.3 [b]	21.9±0.4 [a]	21.5±0.2 [a]	20.7±1.1 [b]
	15	20.6±0.4 [b]	21.0±0.2 [a]	20.7±0.2 [a]	21.9±1.1 [a]
Winter	6	24.5±1.3 [a]	24.2±0.9 [a]	24.6±0.5 [a]	25.1±0.3 [a]
	9	27.7±0.5 [a]	27.5±0.5 [a]	28.3±0.4 [a]	27.8±0.3 [a]
	12	26.3±0.4 [a]	25.6±0.5 [a]	27.9±0.3 [a]	26.3±0.7 [a]
	15	25.7±1.0 [a]	26.7±1.2 [a]	26.1±1.7 [a]	26.6±0.6 [a]
Dry	6	23.5±0.1 [ab]	24.1±0.3 [b]	23.6±0.1 [b]	24.1±0.5 [b]
	9	23.4±0.4 [ab]	22.4±0.5 [b]	21.7±0.3 [b]	21.8±0.5 [b]
	12	20.0±0.7 [b]	31.5±1.5 [a]	26.0±2.3 [ab]	30.8±0.2 [a]
	15	29.1±0.8 [a]	30.9±0.7 [a]	31.0±0.5 [a]	30.2±0.2 [a]

For each accession within season, different letters in the same column mean significative differences ($P \leq 0.0001$).

Table 4. Crude protein content in four accessions of *C. argentea*, at four regrowth ages (averaged over accessions) in the rainy, winter, and dry seasons.

of cutting and age of regrowth did not affect the content of CP, resulting in a small range of values: 191–207 g·kg^{-1} [26].

In the department of Cauca, Colombia (1800 mm annual rainfall), 38 accessions of *C. argentea* were evaluated, including accessions 18516, 18668, and 18676, reporting a range of CP of 184–237 g·kg^{-1} in leaves [27]. These concentrations coincide with the range of values obtained in this experiment.

2.2.3.3. NDF, ADF, and lignin according to season and regrowth age

Mean contents of NDF, ADF, and LIG related to season and age of regrowth are shown in **Table 5**. The responses of these variables to regrowth age were determined by the season. The NDF content at regrowth age from 6 to 12 weeks was similar; and an increase close to 6% units was registered at 15 weeks of regrowth. This variable showed ups and downs during the winter season: at the age of 3 weeks of regrowth, the NDF content was lower, followed by an increase of 7% units in 6 and 9 weeks of regrowth; after that, a decrease around 3% units at 15 weeks of regrowth was recorded. Neither ADF nor LIG increased, as expected, due to the effect of regrowth age pattern. This response is similar to the results found by other authors [27] during the rainy season, where NDF and ADF were lower (42 and 26%) with respect to the dry period (43 and 29%). These results indicated that the climatic season affected the quality of the plants.

Season	Variable (%)	Harvesting age (weeks)			
		6	9	12	15
Rainy	NDF	56.7 ± 1.1^{b}	55.5 ± 0.2^{b}	55.9 ± 0.2^{b}	61.8 ± 1.2^{a}
	ADF	35.7 ± 0.8^{b}	35.5 ± 0.3^{b}	36.9 ± 0.4^{b}	41.8 ± 1.0^{a}
	LIG	14.8 ± 0.8^{b}	18.5 ± 0.2^{a}	18.0 ± 1.0^{ab}	20.6 ± 0.5^{a}
Winter	NDF	58.7 ± 0.7^{c}	65.4 ± 0.9^{a}	65.6 ± 0.5^{a}	62.3 ± 1.1^{b}
	ADF	49.5 ± 0.6^{a}	42.6 ± 0.6^{b}	47.2 ± 1.0^{a}	40.4 ± 0.8^{b}
	LIG	26.0 ± 1.0^{a}	22.6 ± 0.4^{ab}	26.9 ± 1.2^{a}	19.2 ± 0.7^{b}
Dry	NDF	65.5 ± 0.8^{bc}	64.2 ± 0.6^{c}	67.4 ± 1.0^{ab}	69.3 ± 1.0^{a}
	ADF	46.5 ± 0.8^{a}	48.8 ± 0.4^{a}	48.6 ± 0.9^{a}	47.4 ± 1.4^{a}
	LIG	24.3 ± 0.6^{a}	24.7 ± 0.3^{a}	23.3 ± 1.5^{a}	23.8 ± 0.7^{a}

For each regrowth age within season, means in rows followed by different letters differ statistically ($P \leq 0.0001$).

Table 5. Neutral detergent fiber (NDF), acid detergent fiber (ADF), and lignin (LIG) in *Cratylia argentea*, at four harvesting ages, in three climatic seasons (average of the four accessions).

2.2.3.4. Degradation kinetics of dry matter of leaves and stems

In general, the model parameters of degradation [22], namely the rapidly degradable fraction (a), slowly degradable fraction (b), potentially degradable DM (a + b fractions), and the rate of degradation (c), were not affected by accession, week, or their interaction ($P > 0.05$). The parameter 'a' was affected by week, accession × week, and accession in the rainy, winter, and dry seasons, respectively ($P < 0.05$). Parameters "b" and "c" were only affected by accession in the dry season. The parameters (a + b) were similar during the rainy and dry seasons, considering the age of harvest as well as accession. During the winter season a high variation was observed (**Table 6**).

Accession/age	a	b	c	RSD	R^2
18516	15.4–32.6	32.6–39.0	0.04–0.06	2.42–4.82	0.89–0.96
18666	15.8–29.6	33.5–39.9	0.03–0.06	2.14–3.79	0.93–0.96
18668	28.9–31.1	34.7–42.1	0.01–0.06	21.8–4.52	0.89–0.97
18676	25.4–32.6	31.0–37.4	0.04–0.05	2.21–3.84	0.93–0.95
6	28.8–32.6	32.1–42.3	0.04–0.05	2.11–4.33	0.92–0.97
9	23.4–35.4	34.3–35.6	0.03–0.05	2.20–4.63	0.89–0.97
12	20.4–31.0	31.7–36.5	0.04–0.06	2.15–4.40	0.91–0.96
15	21.8–30.5	33.1–45.2	0.02–0.07	2.77–3.59	0.94–0.96

RSD: residual standard deviation.

Table 6. Ranges for three seasons (rainy, winter, and dry) in parameters for the Ørskov equation, of the four *Cratylia argentea* accessions and four regrowth ages, during the rainy season of 2007, obtained as least square means.

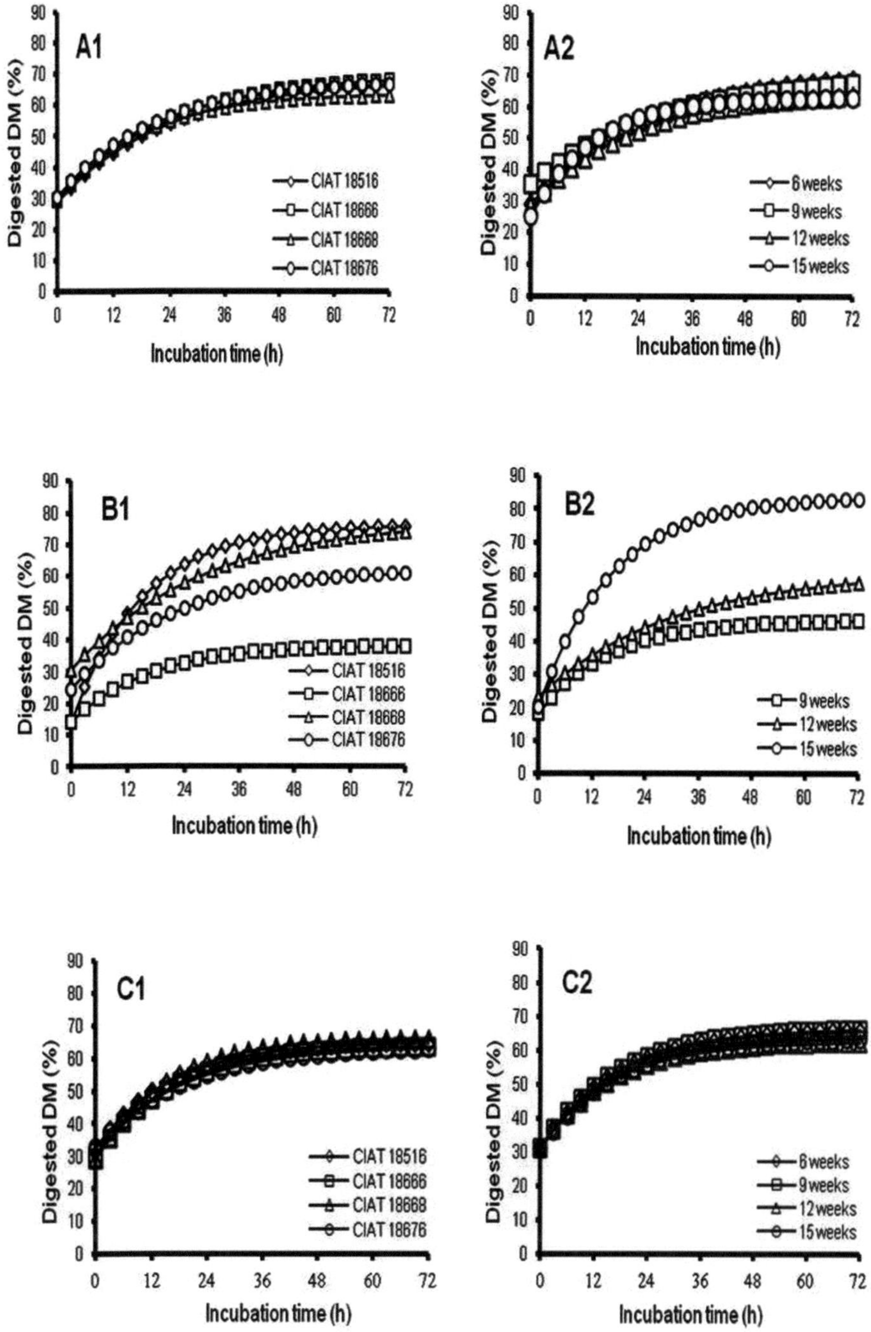

Figure 4. In situ dry matter (DM) degradation (%) of *Cratylia argentea* in three climatic (A = rainy, B = winter, C = dry) seasons, by accessions (A1, B1, C1) and by cutting ages (A2, B2, C2).

A value of 36% was reported by other authors for "a", however, other legume species showed values from 29 to 60% [28]. Other researchers have reported similar values in tropical native woody legumes [15]. Also, degradation rate values (c) coincide with the range of 7–8%, reported by other authors [14].

Figure 4 shows the degradation kinetics of dry matter (leaves + stems < 3 mm) in the rumen, according to the described model [11], for accessions, harvest ages, and ages. During the rainy season, degradation per accession and per week has a very similar pattern, reaching for both cases a value of 66 and 65%, respectively, at 72 h. Considering the age of 9 weeks, a more accelerated degradation was observed during the first 6 h of incubation. A slight variation for accessions and age of regrowth was observed during the winter period. The accessions CIAT 18668 and 18676 highlighted over the rest, but the trend for regrowth age was as expected, and higher digestibility values were presented at 9 ($R^2 = 0.96$) weeks. During the rainy and dry seasons, 48 h in situ DM degradability values for both accessions and regrowth ages were above 60%. Values lower than 35% in leaves of *C. argentea* harvested every 3 months had been reported by other researchers [15].

2.2.4. Conclusions

C. argentea is a reliable forage resource for the dry season in silvopastoral systems, mainly for its high performance in this season. Forage production increased as the ages of regrowth also increased. Considering the obtained results, mainly the quality of evaluated materials, their use at 9–12 weeks of regrowth could be suggested. Also, it is important to emphasize that *C. argentea* has a great potential as a forage resource, observing the high content of CP and digestibility during the rainy and dry seasons.

Author details

Braulio Valles-de la Mora*, Epigmenio Castillo-Gallegos, Jesús Jarillo-Rodríguez, Eliazar Ocaña-Zavaleta and Miguel Angel Alonso-Díaz

*Address all correspondence to: braulio_36@hotmail.com

Facultad de Medicina Veterinaria y Zootecnia, Centro de Enseñanza, Investigación y Extensión en Ganadería Tropical (CEIEGT), Universidad Nacional Autónoma de México, México

References

[1] Bosman HG, Castillo E, Valles B, De Lucía GR. Composición botánica y nodulación de leguminosas en las pasturas nativas de la planicie costera del Golfo de México. Pasturas Tropicales (Colombia). 1990;**12**(1):1-8

[2] Hernández T, Valles B, Castillo E. Evaluación de gramíneas y leguminosas forrajeras en Veracruz, México. Pasturas Tropicales (Colombia). 1990;**12**(3):29-33

[3] Garza TR, Martínez G, Treviño M, Monroy J, Pérez V, Chapa GO. Evaluación de 14 zacates en la región de Hueytamalco, Puebla. Tecnica Pecuaria en México 1973;**24**:7-16. DOI: 10.22319/rmcpv%254107

[4] Pizarro EA, Carvalho MA. Alternative forages for the tropics: Arachis and Paspalum. In: Springer TL, Pittman RN, editors. Identifying Germplasm for Successful Forage Legume-Grass Interactions. Published by The Department of Agriculture, Agriculture Research Service, Washington, D.C.; 1996. pp. 1-14

[5] Abberton MT, MacDuff JH, Marshall AH, Humphreys MW. The genetic improvement of forage grasses and legumes to reduce greenhouse gas emissions. 2007. Available from: http://www.fao.org/ag/agp/agpc/doc/climatechange/papers/abberton_%20geneticimprovement.pdf [Accessed: April 14, 2011]

[6] Castillo GE, Valles MB, Mannetje Lt, Aluja-Schunemann A. Efecto de introducir *Arachis pintoi* sobre variables del suelo de pasturas de grama nativa del trópico húmedo mexicano. Tecnica Pecuaria en México. 2005;**43**(2):287-295. DOI: 10.22319/rmcpv%251372

[7] Enríquez QJF, Hernández A, Pérez J, Quero A, Moreno J. Densidad de siembra y frecuencias de corte en el rendimiento de *Cratylia argentea* (Desvaux) O. Kuntze en el sur de Veracruz. Tecnica Pecuaria en México. 2003;**42**(1):75-84. DOI: 10.22319/rmcpv%251285

[8] Lascano C, Rincón A, Plazas C, Ávila P, Bueno G, Argel P. Cultivar Veranera (Cratylia argentea Desvaux O. Kuntze)—Leguminosa arbustiva de usos múltiples para zonas con periodos prolongados de sequía en Colombia. Cali, Colombia: Corpoica, CIAT; 2005. p. 27

[9] AOAC. Official Methods of Analysis. 13th ed. Mensha, Wisconsin: Association of Official Analytical Chemists; 1984. p. 1545

[10] Van Soest PJ, Robertson JB, Lewis BA. Methods for dietary fiber, neutral detergent fiber, and non-starch polysaccharides in relation to animal nutrition. Journal of Dairy Science. 1991;**74**:473-481. DOI: 10.3168/jds.S0022-0302(91)78551-2

[11] Ørskov ER, McDonald I. The estimation of protein degradability in the rumen from incubation measurements weighted according to rate of passage. Journal of Agricultural Science. 1979;**92**:499-503. DOI: 10.1017/S0021859600063048

[12] Menke KH, Raab L, Salewski A, Steingass H, Fritz D, Schneider W. The estimation of the digestibility and metabolizable energy content of ruminant feeding stuffs from the gas production when they are incubated with rumen liquor *in vitro*. Journal of Agricultural Science. 1979;**93**:217-222. DOI: 10.1017/S0021859600086305

[13] SAS Institute Inc. SAS/STAT® 9.22 User's Guide: The GLM Procedure (Book Excerpt). Chapter 39. Cary, NC: SAS Institute Inc.; 2010. pp. 2986-3177

[14] Krishnamoorthy U, Soller H, Steingass H, Menke KH. A comparative study on rumen fermentation of energy supplements *in vitro*. Journal of Animal Physiology and Animal Nutrition. 1991;**65**:28. DOI: 10.1111/j.1439-0396.1991.tb00237.x

[15] Suárez JC, Ramírez BL, Velásquez JE. Comportamiento agronómico de cinco especies forrajeras bajo el sistema de corte y acarreo en suelos de terraza y mesón en el piedemonte amazónico colombiano. Zootecnia Tropical. 2008;**26**:347-350

[16] Franco VMH. Evaluación de la calidad nutricional de *Cratylia argentea* como suplemento en el sistema de producción de doble propósito en el trópico subhúmedo de Costa Rica. Tesis de Magister Scientiae. Turrialba, Costa Rica: Catie; 1997

[17] Barahona R, Lascano CE, Narvaez N, Owen E, Morris P, Theodorou MK. *In vitro* degradability of mature and immature leaves of tropical forage legumes differing in condensed tannin and non-starch polysaccharide content and composition. Journal of the Science of Food and Agriculture. 2003;**83**:1256-1266. DOI: 10.1002/jsfa.1534

[18] Bernal L, Ávila P, Ramírez G, Lascano CE, Tiemann T, Hess H. Efecto del ensilaje y el heno de *Calliandra calothyrsus, Flemingia macrophylla, Cratylia argentea* y *Vigna unguiculata* sobre la producción de gas *in vitro*. Archivos Latinoamericanos de Producción Animal. 2008;**16**(3):101-107

[19] Casanova-Lugo F, Petit-Aldana J, Solorio-Sanchez FJ, Parsons D, Ramírez-Avilés L. Forage yield and quality of *Leucaena leucocephala* and *Guazuma ulmifolia* in mixed and pure fodder banks systems in Yucatan, Mexico. Agroforestry Systems. 2014;**88**:29-39. DOI: 10.1007/s10457-013-9652-7

[20] Sosa RE, Cabrera TE, Pérez RD, Ortega RL. Producción estacional de materia seca de gramíneas y leguminosas forrajeras con cortes en el estado de Quintana Roo. Técnica Pecuaria en México. 2008;**46**:413-426. DOI: 10.22319/rmcpv%251795

[21] Toledo MJ, Schultze-Kraft R. Metodología para la evaluación agronómica de pastos tropicales. In: Toledo JM, editor. Manual para la evaluación agronómica. Cali, Colombia: Red Internacional de Evaluación de Pastos Tropicales-Centro Internacional de Agricultura Tropical (CIAT); 1982. pp. 91-111

[22] Ørskov ER, De Hovell BFD, Mould F. The use of the nylon bag technique for the evaluation of feedstuffs. Tropical Animal Production. 1980;**5**:195-213. Available from: http://www.fao.org/ag/aga/agap/frg/tap53/53_1.pdf [ISSN: 0250-5576]

[23] Statistical Analysis System (SAS). SAS/STAT User's Guide. Release 8.0 Edition. Cary, USA: Autor; 1999

[24] Rodríguez PI, Guevara DE. Producción de materia seca y valor nutritivo de la leguminosa arbustiva *Cratylia argentea* en el sur del estado Anzoátegui, Venezuela. Revista Científica FCV-LUZ. 2002;**12**(2):589-594. Available from: http://www.saber.ula.ve/revistacientifica/n12/pdfs/articulo_56.pdf

[25] Santana RMO, Medina SM. Producción de materia seca y calidad forrajera de *Cratylia argentea* (Desv.) O. Kuntze bajo tres alturas y edades de corte en bosque húmedo tropical. Livestock Research for Rural Development. 2005;**17**(10). Available from: http://www.lrrd.org/lrrd17/10/sant17116.htm [Accessed: February 17, 2010]

[26] Andersson MS, Peters M, Schultze-Kraft R, Franco LH, Lascano CE. Phenological, agronomic and forage quality diversity among germplasm accessions of the tropical legume shrub *Cratylia argentea*. Journal of Agricultural Science. 2006;**144**:237-248. DOI: 10.1017/S0021859606006034

[27] Flores OI, Bolivar DM, Botero JA, Ibrahim MA. Parámetros nutricionales de algunas arbóreas leguminosas y no leguminosas con potencial forrajera para la suplementación de ruminantes en el trópico. Livestock Research for Rural Development. 1998;**10**:1-5. Available from: http://www.lrrd.org/lrrd10/1/cati101.htm

[28] Cecconello CG, Benezra SM, Obispo EN. Composición química y degradabilidad ruminal de los frutos de algunas especies forrajeras leñosas en un bosque seco tropical. Zootecnia Tropical. 2003;**21**:149-165. Available from: http://www.scielo.org.ve/scielo.php?pid=S0798-72692003000200004&script=sci_arttext

Chapter 4

Halophytes as Forages

Salah A. Attia-Ismail

Additional information is available at the end of the chapter

http://dx.doi.org/10.5772/intechopen.69616

Abstract

It is the chemical composition of the halophyte forages and the digestion process of these forages that matter. As the science gets more advanced and the information about these two points becomes clearer, the view of this information might modify our understanding to these processes. Then, some topics might be dropped, and others might be raised or become more obvious. However, the feeding of halophyte forages as per se has several drawbacks and therefore, they have to be fed in mixed rations, fortifying these rations with energy supplements.

Keywords: halophytes, forages, ruminants, feeding, nutritional values, plant secondary metabolites, protein, energy, rumen function, feed processing

1. Introduction

Halophytes are not a distinct taxonomic group. Halophytes are several species of trees, shrubs, forbs and grasses. They fall into various taxonomic groups, and their life form spectrum exhibits a wide range of variation. When slat tolerant plants are included, the number of halophytes increases significantly. It was estimated [1] that the flowering plants are about to be 350 families of which one-third is halophyte forages. It was found [2] that 50% of the genera belong to 20 of these families. It is concluded, then, that the halophyte forages do not constitute a family per se but they are widely distributed within different families of flowering plants. The fact that the limited number of halophytic species is spread among so many different families indicates that halophytism, even though a trait controlled by several genes, is not such a complex characteristic that only arose once during evolution. The word halophyte, then, does not imply any reference to being a particular taxon or any specific geographic or physiogeographic area [3].

Nature and ecology of halophytes are very complex [4]. They do not necessarily need salinity to grow. Halophytes survive salt concentrations around 200 mM NaCl or more in order to reproduce in environments where they constitute about 1% of the world's flora [5].

It is estimated that 7–10% of the world land area is salt affected [6]. Salt-affected soils happen to occur in all over the world and almost under all climatic conditions. Their distribution, however, is relatively more extensive in the arid and semi-arid regions compared to the humid regions.

The natural resources in Egypt have been diminishing because of increased demands. The increased population and the decline of the arable lands make it inevitable to utilize marginal and long-neglected natural resources and re-assess them in preparation for utilization. Halophyte plants are widely distributed throughout several regions of Egypt due to the presence of numerous saline areas along the Mediterranean Sea and Red Sea shores and inlands (littoral salt marshes and inland salt marshes). The less and unpalatable plant species represent approximately 70% of the total coverage. FAO [7] has estimated that salt-affected soil area in Egypt is about 7360 (ha). The arid climate of Egypt is characterized by high evaporation rates (1500–2400 mm/year), and a little rainfall (5–200 mm/year), which may add up to the existing salt affected soils.

Main causes of salinity development are irrigation with saline water; disturbance of the water balance between rainfall, on the one hand, and streamflow, groundwater level, and evapotranspiration, on the other; overgrazing, and cutting bushes; water percolation through saline materials; and intrusion of seawater [8].

2. Production of green biomass from halophytes

Halophytes can grow naturally or be planted. The biomass production and quality of the natural vegetation of halophytes in such areas vary considerably from season to season and from area to area depending on several factors, mainly environmental ones. In almost any forage populations, of a given species of a browse, there are various degrees of palatability from one plant to the other.

Suppressed growth of field crops is a direct result of the presence of salt in soils and the irrigation with saline waters. Therefore, the yield of these crops is affected dramatically where the expected yield relates to the plant species and salt concentrations either in soil or irrigation water. The studies to estimate the yield potential of halophyte forages were carried out on a laboratory scale. Very few studies were performed in the field. It was found that some halophyte forages like some species of *Atriplex* (e.g., *A. nummularia, A. griffithii* and *A. hortensis*) could tolerate high concentration of salt. It was found [9] that optimal growth of such species would be at 5–10 g/l^{-1} NaCl. The estimated yield value of *A. leucoclada* in the high salinity experimental site was 3735 kg fresh weight and 2058 kg of dry weight [10]. Some species of *Atriplex* yielded 1.26–2.09 kg/m^2 dry matter, 15.5–39.5% crude fiber and 10.2–19.5% crude protein [11].

Kochia indica was found [12] to produce fresh biomass of 8.5 kg per bush from March through August in India. **Table 1** represents some information gathered [13] concerning the yield of

Plant species	Salt concentration (mM)	Yield (kg m^{-2} $year^{-1}$)
Aster tripolium	40	14.0 (fresh weight basis)
Atriplex lentiformis	500	1.8 (dry weight basis)
Atriplex triangularis	150	21.3 (fresh weight basis)
Batis maritima	500	1.7 (dry weight basis)
Salicornia europaea	500	1.5 (dry weight basis)
Salicornia persica	100	15.0 (fresh weight basis)
Sarcocornia fruticosa	100	28.0 (fresh weight basis)

Table 1. Yields obtained from halophyte crops grown under field conditions [13].

some halophytic forages grown under high salt effects. However, the estimated yield of halophytic forages reaches about 4–5 billion tons [14] resulting from 450 million hectares in the world according to FAO.

3. Feeding and nutritional value of halophytes

3.1. Quality as animal feed components

The quality might be the extent to which a halophytic or salt tolerant plant, as forage, has the potentiality to reach the required animal response. The quality of halophytes as forage varies greatly among and within each crop. In order to determine forage quality, different issues have to be taken into consideration.

The factors that affect forage quality include palatability, nutrient contents (chemical compositions), plant secondary metabolites [15], feeding value (voluntary animal intake, nutrient digestibility), and eventually animal performance.

Analyzing forages for nutrient content (chemical compositions) can be used to determine the quality of forage if it is adequate to meet the animal requirements and to be used for proper ration supplementation. Limitations [16] of halophytic forages as feeds for animals (i.e., accounting for non-protein nitrogen and non-nutritional components) could represent a problem in formulating rations. He also referred to the palatability issues of the halophytic forages as important factors in determining the acceptability of these forages by animals and to which extent they might be consumed. The other factors that assess the quality of these forages (like an assessment of feeding and nutritional values) might be looked upon after the issues of palatability are addressed.

3.2. Palatability and preference

The definition of palatability has been an argument. Regardless of the scientific controversy over this issue, the most agreed upon is that palatability of a feed is the ration between the consumed and offered amounts of feed by any class of herbivores animals on a given time [17, 18]. The palatability and feeding values of individual halophytes or any other types of rangelands vary widely from virtually zero to very high. In almost any forage populations, of a given species of a browse, there are various degrees of palatability from one plant to the other. Palatability depends (among other factors) on the relative abundance of the species on the rangeland. Considering all other conditions being equal, the palatability of a given plant is inversely related to its profusion on the range.

Regardless of the plant internal factors, animal factors also govern the palatability of the halophyte forages. These factors may include, but not limited to, animal species and race, age, physiological state and health status, feeding habits, animal conditions as controlled by nutrition.

Chemical compositions of halophyte forages affect also their palatability. For instance, if the crude fiber percentage is high in forage, it will play an important role in its selection by livestock. Forages with high fiber content are usually better accepted by cattle than by sheep and goats. Mineral content [19, 20] in low rain fall areas compared to high rain ones, the ash percentage (when silica-free minerals are concerned) in halophyte forages could be a critical factor to the palatability, may be because of dilution rate. **Table 2** shows the palatability of halophytic plants for different animal species.

3.3. Chemical compositions

Halophytic plant species vary considerably in their chemical composition, nutritive value, and palatability. The chemical composition of any animal feed is the first indicator if its nutritional value to the animals is considered. Nutritive value is first determined by nutrient concentration through the determination of the feed plant chemical composition. The differences in chemical compositions, and hence nutrient contents, of halophytic forages, may be related to the variations in factors that control plant growth (e.g., soil fertility, soil salinity, environmental factors like rain and temperature, etc.). Therefore, the determination of nutrient contents of these forages is a must to assess their quality as feed components.

3.3.1. Ash contents and mineral compositions

The fact that a high content of ash is a typical characteristic of halophytic forages has resulted in divisive concerns over the bioavailability of mineral contents of these forages. The concerns about this issue are justifiable since the raised questions were to what extent this could affect the nutritional value of these types of forages, how much the mineral contents of halophytic forages could satisfy these requirements and whether they poisonous, in case if they exceed the animal requirements.

However, the mineral profiles of halophytic forages differ from those of traditional ones. These differences may due in part to [19] forage species, stage of growth, seasonality, the degree of soil and water salinity, etc. The concentrations of some mineral contents of halophytic forages are shown in **Table 3**. It appears that these forages could be a source of some minerals to meet ruminant animal requirements. In this context, the concentrations of these minerals may balance the deficiency that may result from in areas depending on grazing ranges (e.g., desert and coastal areas).

Animal species	Plant species
Sheep, goats	*Alhagi maurorum*
Camels	*Arhthrocenemon glaucum*
All species	*Atriplex halimus*
Sheep, goats	*Atriplex leucoclada*
All species	*Atriplex nummularia*
Camels	*Halocnemom strobilaceum*
Nil	*Haloxylon salicornicum*
All species	*Juncus acutus*
All species	*Nitraria retusa*
Camels	*Salicornia fruticosa*
All species	*Salsola tetrandra*
All species	*Suaeda fruticosa*
All species	*Limoniastrum monopetalum*
Goats, camels	*Tamarix aphylla*
All species	*Tamarix mannifera*
Nil	*Zygophyllum album*
Camels	*Zygophyllum simplex*
Camels, goats	*Zygophyllum decumbens*

Table 2. Palatability of some halophytic plants for different animal species [21].

The aspects of ash contents and mineral compositions of halophytes are discussed in detail by [19, 20]. The mineral profiles of some halophytic forages in Australia were examined [23]. The authors found that some ions are present in frequent patterns, especially in certain taxons. Sodium salts (especially chlorides) were found to accumulate in large concentrations in dicotyledons compared with sulfate salts. Chenopodianceae and Caryphyllaceae were found to have normal concentrations of free oxalates. Other dicotyledons found to have moderate salt contents. The ratio of K:Na in these plants was found to be less than one. They also found that the patterns of mineral salts in monocotyledons were in contrast to those of dicotyledons. Low salt concentrations are characteristic monocotyledons like Poaceae and that the K:Na ration is more than one. Similar results, later on, were found [24] in Wales.

All halophyte forage species contain adequate amounts of major and minor minerals (**Table 3**) to apparently meet the mineral requirements of ruminants except for both of phosphorus and sulfur according to [25].

The high levels of mineral contents of halophyte forages do not exceed the normal levels of the requirements of livestock, especially ruminants. However, it is preferred to include supplements of trace and minor mineral in diets in order to correct for any deficiency that may occur.

	Ash (%)	Ca (%)	P (%)	Na (%)	K (%)	Mg (%)	S (%)	Zn (ppm)	Cu (ppm)	Fe (ppm)	Mn (ppm)
Acacia saligna	8.83	3.75	–	1.15	1.05	6.14	–	140.5	–	–	–
Atriplex nummularia	18.91	2.08	1.17	4.99	2.99	15.63	–	133.5	60.52	–	–
Atriplex halimus	29.20	1.69	0.32	3.91	0.57	0.32	0.17	64	10	503	51
Nitraria retusa	9.99	1.96	0.22	5.35	0.66	0.36	0.14	32	11	567	62
Tamarix mannifera	8.06	3.01	0.01	2.70	0.91	0.46	0.09	45	16	291	52
Suaeda fruticosa	30.2	2.11	0.41	4.06	1.29	0.30	0.20	55	13	674	88
Salsola tetrandra	12.9	3.98	0.16	5.65	1.45	0.59	0.12	44	8.88	664	79
Zygophyllum album	34.84	2.26	0.14	2.89	1.14	0.64	0.09	41	7.78	393	52

Table 3. Overall average values of some mineral composition in halophytic plants (DM basis) grown in Sinai and the North Western coast of Egypt (adapted from [19–22]).

3.3.2. Protein and amino acid contents

It has been long recognized that environmental conditions play a major role in determining the quantity and quality of nutrients produced by halophytes. It is reported that proteins level decreased under salinity is due to low uptake of nitrate ions [26] and due to other factors.

The biochemical processes that take place within halophytic forages for the biosynthesis of different nutrients seem to be affected by the high concentrations of salts [27]. These processes include the protein and amino acid formation [28]. The increases in salt concentrations cause decreases in the protein synthesis and its hydrolysis as well [29]. This process results in the production of amino acids in some halophytic forages. The antagonistic effect of increased salinity on protein synthesis is, then, clear. However, some amino acids like aspartate and glutamate play a critical role in the adaptation of halophytic forages to salt stress. Concentrations of aspartate, glutamate, glycine, histidine, lysine, and arginine amino acids were found to increase as the salinity levels increase [30]. Within the salt-tolerant sorghum types, protein content decreases as the salinity increases leading to the increase of non-protein-nitrogen [31]. It seems, therefore, that with a decrease in soil salinity, the available nitrogen increases significantly.

In general, the nitrogen contents of most of the halophytic forages are reasonable and appear to cover the requirements of grazing animals. As mentioned above, most of the nitrogen contents of halophytic forages are in the form of amino acids (NPN). It was found [32] that almost 42% of nitrogen contents in *Atriplex barclayana* were in the form NPN. This has certain implications in animal nutrition, as an available energy source should be included in the rations of animals feeding on halophytic forages. This inclusion may have its impact on the utilization and efficiency of nitrogen digestion [33].

In evaluating proteins present as a dietary nutrient in halophytes, one should take several issues into consideration. First is the high percent of non-protein nitrogen portion of the crude protein content. Second consideration is that the increased solubility of proteins contents of halophytes arises from their presence as leaf proteins (leaf proteins are usually highly soluble) and because halophytes react in different mechanisms to high salt stress. Halophytes store most of their proteins in the leaves at the beginning and later on (after plant maturation) in the seeds (**Table 4**). The third consideration results from the high solubility of proteins. This characteristic of leaf protein has its implication on their degradability by ruminal microorganisms which tend to be high. The rumen microflora act on dietary soluble proteins once ingested. They degrade them in order to build their body protein. If a readily available energy source is lacking during this process, the degraded protein is, then, wasted and the animal does not get benefit out of it. The literature on halophytes shows that the digestibility of crude fat contents (or ether extract) is low. They also have low contents of soluble carbohydrates. This leads to decreased synthesis of microbial proteins in the rumen of animals. Protein supply to the animal is not, then, sufficient to meet its requirements of proteins even at maintenance level. That is why animals feeding on halophytes alone loss weight. The supplementation of a readily available carbohydrate source is a must in this case in order to increase the synthesis of microbial proteins. The coincidence of the release of both degraded soluble proteins and the highly soluble carbohydrates is a critical process. The non-degraded cereal proteins provide

Halophytic plants	Plant part and/or maturity stage	N (%)	Protein (%)
Acacia saligna	Whole	2.21	13.8
Atriplex halimus	Whole plant	2.11	13.1875
Atriplex nummularia	Whole plant	2.03	12.6875
	Fruits	1.65	10.3125
Salsola tetrandra	Whole plant	1.08	6.75
Suaeda foliosa	Leaves	2.67	16.6875
	Stem	2.69	16.8125
Suaeda fruticosa	Whole plant	1.94	12.125
Tamarix mannifera	Whole plant	1.22	7.625
Zygophyllum album	Whole plant	1.05	6.5625

Table 4. Nitrogen and crude protein contents of different parts of some world halophytes [14].

the animal with a source of protected protein, hence, providing the animal with true proteins. All these together may explain the positive response of animal fed halophytes when supplemented with energy concentrate. It is, then, necessary or may be vital to supplement animals fed on halophytes with cereal grain energy supplement.

3.3.3. Energy contents

The definition of feed gross energy (GE) is the total combustion heat of any feed substance expressed in calories or joules per unit of dry matter. The digestible energy (DE) is the amount of gross energy minus the energy lost in feces, while the metabolizable energy (ME) is the digestible energy minus the amount of energy lost in urine and gasses. The net energy (NE) for maintenance is the metabolizable energy minus that lost as heat. The most common energy form used to express the energy contents of halophytic forages is metabolizable energy.

However, the reported energy content of halophytes is usually estimated in vitro. These values may be unrealistic ones and do not represent real values of in vivo values. However, these *in vitro* values relate to some extent to the *in vivo* ones. **Table 5** was compiled [33] to show the inconsistency of *in vitro* values compared to those produced in vivo.

However, the nutritive value of halophyte species such as metabolizable energy (ME) appears to depend strongly on plant maturity. Energy contents of both traditional forages and halophytic ones (**Table 6**) were found to be similar and had no significant differences. The question is, then, is there a difference in the efficiency by which the energy is utilized in both types of forages? The published values are contradicting. When *A. nummularia* hay was compared with alfalfa hay [34], the ME intake was not different. Coastal grasses like Aeluropus lagopoides and Sporobolus tremulus appeared to have adequate energy contents [35] to meet the maintenance requirements of beef cattle, while those grazing animals on *A. nummularia* need energy supplementation than any other supplementation [36]. It was concluded [37] that the low nutrient digestion and utilization of halophyte forages could be attributable to the low energy contents.

	In vivo	Pepsin-cellulase	Pepsin-cellulase corrected*	NIRS
Sample 1	58			76
Sample 2	52	77	70	
Sample 3	45	77	71	

* Corrected with non-halophyte calibration.

Table 5. Estimates of *in vitro* and *in vivo* of DOMD values (adapted from [33].

Halophytic forage	ME (Mcal/kg)	conventional forages	ME (Mcal/kg)
Aeluropus lagopoides[1]	2.30	*Medicago sativa*[3]	2.20
Sporobolus tremulus[1]	2.38	*Cynodon dactylon*[3]	2.49
Paspalum paspalodes[1]	2.53	*Sorghum vulgare*[3]	1.75
Paspalidium geminatum[1]	2.33	*Zea mays*[3]	2.97
Atriplex nummularia[2]	2.82	*Trifolium alexandrinum*[3]	1.99
Salsola tragus[2]	2.56	*Lolium multiflorum*[3]	2.50

Table 6. Examples of Digestible Energy (DE) and Metabolizable Energy (ME) values of some halophytes compared to some traditional forages [35, 38, 39].

4. Effect of feeding halophytic forages on rumen function

The microbial population in the rumen and its metabolism is anticipated to be affected by the salt load which increases the osmotic pressure [40–42]. The elevated osmotic pressure within the rumen environment is assumed to be critical to the protozoa growth. This may increase the outflow rate and, hence, decrease the protozoa population [43]. Artificial raises [44] in the osmotic pressure of the rumen up to 400 mOsmol/kg and found that the cellulose digestion was inhibited. The increased flow rate due to the increased salt load in the rumen depressed the protozoal population [37]. On the other hand, Ref. [45] found a significant increase in protozoal count ($\times 10^3$/ml rumen fluid) when camels were fed ration containing *A. nummularia* compared with those fed Acacia saligna and treated rice straw rations. The same increments in the ruminal protozoal population were found when camels were fed on berseem hay compared to those fed traditional rations. It seems that the increased load of salts in the rumen as a result of feeding desert halophytic forages imposes ion burden that needs to be buffered. Therefore, ruminants fed rations containing halophytes are anticipated to release more saliva and may have elevated pH values than those fed grains [46].

5. Limitations of feeding halophytes to ruminants

The low intake of fresh and air-dried halophytic species could be attributed to several factors: (1) high Na, Ca and silica contents, (2) higher levels of ADL and NDF and (3) many shrubs contain higher levels of plant secondary metabolites, (4) low energy contents, (5) low crude protein contents and (6) high percentage of non-protein nitrogen. Nutrient detergent fiber (NDF) is a good indicator for forage intake. The low NDF value (50.2%) for the fresh Potamogeton *crispus* would explain the higher intake by sheep compared to fresh *Tamarix mannifera* and *Glinus iotoides* [47]. The limited halophytic intake and digestion may be attributable to the low crude protein contents (around 6%) and greater levels of NDF, ADF, and ADL. This case is well illustrated in *T. mannifera* and *G. iotoides*. The *P. crispus* showed opposing trend. When *P. crispus* fed to sheep, the TDN and DCP values were high, and the animals were in positive nitrogen balance.

Voluntary feed intake and nutrient digestion/unit of feed are the criteria against which the feeding value of feeds is considered. However, factors like physical and chemical properties of halophytes that are used to defend the plants against predators may considerably limit the feeding values of such forages. Physical factors like the presence of spines and thrones may include the so-called barbed-wire syndrome [48]. Chemical factors may include the higher salinity, silica, and fiber. The presence of lignin and the degree of lignification also affect the nutritive value of halophytes as animal feed components. The secondary plant metabolites that limit the feeding value of halophytes are another example of the chemical defense of halophytes. Salt load present in halophytes affects their palatability and acceptability as well and, therefore, the intake [49, 50].

The limited halophytic intake and digestion may be attributable to the low crude protein contents (around 6%) and greater levels of NDF, ADF, and ADL. This case is well illustrated in *T. mannifera* and *G. iotoides*. The *P. crispus* showed opposing trend. When *P. crispus* fed to sheep, the TDN and DCP values were high and the animals were in positive nitrogen balance.

Voluntary feed intake and nutrient digestion/unit of feed are the criteria against which the feeding value of feeds are considered. However, factors like physical and chemical properties of halophytes that are used to defend the plants against predators may considerably limit the feeding values of such forages. Physical factors like the presence of spines and thrones may include the so-called barbed-wire syndrome [48]. Chemical factors may include the higher salinity, silica, and fiber. The presence of lignin and the degree of lignification also affect the nutritive value of halophytes as animal feed components. The secondary plant metabolites that limit the feeding value of halophytes are another example of the chemical defense of halophytes. Salt load present in halophytes affects their palatability and acceptability as well and, therefore, the intake [49, 50]. Animals make selection and palatability on basis of their acceptability to the halophytic plant [16] gives more detailed attention to how to assess the nutritive value of halophytes).

Some halophytes are toxic [41]. **Table 7** shows a screening of anti-nutritional factors present in some halophytes. The toxicity results from several secondary metabolites in the plants. However, the rate of toxicity is affected by several factors such as rate of ingestion, type, and rate of microbial transformation of such metabolites in the rumen, rate of gastro-intestinal absorption, liver and kidney enzymatic activity. Alkaloids, saponins, tannins and nitrates are present in most halophytes. High concentrations of alkaloids decrease animal performance and increase diarrhea. Tannins reduce feed intake through reducing palatability resulting from the precipitation that occurs upon reaction of tannins with salivary proteins. Tannins also inhibit digestive enzymes [51].

Fodder crops	Anti-nutritional factors
Atriplex nummularia	Saponin, alkaloids, tanins, nitrate
Atriplex leucoclada	Saponin, alkaloids, tannins
Atriplex halimus	Saponin, flavonoids, alkaloids, tannins, nitrate
Diplache fusca	Flavonoids, alkaloids
Halocnemum strobilaceum	Saponin, flavonoids, alkaloids, tannins, nitrate
Haloxylon salicornicum	Saponin, flavonoids, alkaloids, tannins
Kochia eriophora	Alkaloids, tannins
Juncus acutus	Flavonoids, alkaloids, tannins, nitrate
J. arabicus	Alkaloids, tannins
J. subulatus	Alkaloids, tannins, flavonoids
Limonium pruinosum	Saponin, alkaloids, tannins
Nitraria retusa	Saponin, tannins
Salsola glauco	Saponin, flavonoids, alkaloids
Suaeda fruticosa	Alkaloids, tanins, nitrate
Tamarix aphylla	Saponin, tanins
Salsola tetrandra	Nitrate
Tamarix mannifera	Saponin, tannins
Zygophyllum album	Saponin, flavonoids, alkaloids, tannins, nitrate
Sesbania sesban	Saponin, alkaloids

Table 7. Examples of plant secondary metabolites in halophytic forages [52].

6. Overcoming constraints of halophytes as animal feed

Natural resources have been diminishing because of increased human pressure. This pressure results from the ever-increasing population of the world. Inevitably, under current and predicted future conditions marginal resources and long-neglected natural resources such as halophytic plants have to be re-assessed in preparation for future utilization. Shortage of animal fodder is one of the main constraints of indigenous animal production on salt affected soils of arid and semi-arid regions and limits its expansion. Animal husbandry, as the main income resource for nomads, is based mostly on the natural vegetation for feeding sheep, goats and other herbivores.

The way in which halophytes are used depends very much on the nature of the community that dominates their ecosystem. Evaluation of the possible contribution of halophytes to the economic well-being of the local nomadic communities depends on the understanding of the economy, agrobiology, and ecology of the forage plants and the knowledge of the carrying capacity of the grazing animals. Halophytic plants have long been ignored and viewed as marginal resources.

The use of halophytes for animal feed has several constraints that must be dealt with, on a rational exploratory and experimental basis. The high content of mineral ash, the presence of plant secondary metabolites and the low nitrogen content are examples of the constraints that face animal nutritionists. Little was done in the exploration of the richness of various halophytic species for the purpose of selection of halophytes of high quality for grazing.

Most of the halophytes contain secondary metabolites (tannins, glucosides, flavonoids, alkaloids, terpenoids, cyanides, coumarin, nitrate, oxalate and organic acids). There are many plants capable of producing toxic metabolites including palatable plants [53]. For example, *Nitraria retusa* one of the most palatable grazed halophytic shrub in Egypt contained different proportions of crude alkaloids, saponins in addition to tannins and sterols [54].

Harmful effects of plant secondary metabolites cause great economic losses to livestock producers. However, ruminants are more tolerant to poisonous plants than non-ruminants. Even among ruminants, there are striking differences in tolerance of plant toxicants. In ruminants, tolerance of poisonous plants may be modified by microbial fermentation of ingesta in the reticulorumen, which can diminish toxicity of some plants compounds and increase the toxicity of other. Some plant compounds may be biotransformed within tissues of the host ruminant yielding products that are more toxic or less toxic than the plant compound ingested [55]. Ruminants may convert a toxic substance to another toxic one (cyanide to thiocyanate, which is goitrogenic) [56]. They also may detoxify some substances with a concurrent loss of some nutrients [57]. Methods of overcoming these constraints may include cooking, germination. The effectiveness of these methods differs from one another. On the other hand, some methods (like steam treatment) may improve the nutritive values of halophytes by increasing the accessibility of nutrients. Steam can break down plant secondary metabolites to some extent and may make fat more available [58].

A summary of the plant secondary metabolites, their impacts on animals and some ways to lessen their effects on animals is present in **Table 8**.

Plant secondary metabolite	Impact on animal	Methods to relief
Phenolic compounds	Affect rumen fermentation	**1.** PEG **2.** Physical treatment **3.** Silage
Glycosides: *1. Saponins*	**1.** Bloat **2.** inhibit microbial fermentation **3.** Formation of calcium salt **4.** Decrease growth rate	**1.** Repeated washing with water **2.** Ensiling or wilting in the field
2. Cyanogens	Animal death due to its harmful on hemoglobin	**1.** Add methionine to animal diet (sulfur combines with cyanide to form thiocyanate (non-toxic) **2.** Sun drying

Plant secondary metabolite	Impact on animal	Methods to relief
3. Goitergens	1. Enlargement of thyroid gland 2. Rapid decline in serum thyroxine, 3. Decreased intake 4. Prolonged feeding has produced hair loss, excessive salivation and esophageal lesions	Broken down in the rumen by rumen bacteria
Alkaloids	1. Ataxia 2. Diarrhea 3. Decrease animal performance	1. Air drying 2. Ensiling
Nitrates	1. inhibition of cellulose digestion 2. Combines with hemoglobin, thus reducing the oxygen 3. High nitrates cause abortion in livestock	1. Add grains and vitamin A to the diet 2. Mechanical treatment 3. Add more soluble CHO to increase microbial nitrogen requirements
Oxalate	1. Excess oxalate may result in fatal intoxication with hypocalcaemia, metabolic disturbances and kidney failure 2. May result in fatal intoxication with hypocalcaemia, metabolic disturbances and kidney failure 3. Kidney failure due to the accumulation of oxalate crystals	Animal adaptation because rumen bacteria can degrade it
Phytates	Hypomagnesima (low WBC) Milk fever (decreased Ca & P)	1. Mineral balance 2. Vitamin D injection
Tannins	1. Reduced voluntary feed intake 2. Reduced digestibility of protein and carbohydrate through the inhibition of digestive enzymes 3. May reduce bacterial enzymes 4. Tannins/protein complex that survives in the ruminal environmental may not be digested in the lower tract	Add PEG

Table 8. Plant secondary metabolites and their impact on animals and how to reduce their effects [59].

7. Processing as animal feeds at farm level

Halophytes and salt tolerant fodders contain some physical and salt materials that limit and constrain its palatability and utilization by animals [14]. The main constraints are high ash content (minerals), high fiber content (in particular the lignin and hemicellulose), low protein and energy contents, high presence of secondary metabolites (anti-nutritional factors) produced by plants, such as tannins, etc. which have a direct impact on the processes of digestion in animals. Halophytes and slightly salt tolerant fodders also contain some physical and chemical materials that limit and constrain its palatability and utilization.

Current methods of processing dry forages include chopping, grinding, shredding, silage, feed cubes, hay or mix components in a TMR [60].

Forages are subject to waste when fed directly to livestock. Waste occurs because of livestock discriminately select specific components of forage (leaves, smaller stems), animal trampling, spoiling (urine and manure deposition) or bedding on excess forage.

Processing halophyte forages (whether cultivated or naturally grown) provides some advantages. Processing can maximize the use of forages to be included in a total mixed ration for livestock diets. Processing also ensures livestock diet consistency in a uniform blend. Processing can decrease waste from animal selection and allow more precise ration formulation. Processing benefits include reduced feed waste and the ability to mix diets more precisely with a wider variety of feedstuffs includes reduced feed waste and the ability to mix diets more precisely with a wider variety of feedstuffs. Processing forages will decrease particle size, reduce opportunity for sorting of forages by animals. Processing also can help producers develop more precise and cost-effective rations.

However, the primary benefits of processing will not improve hay quality; however, it potentially can increase DMI within a blended TMR due to a smaller particle size. These benefits need to be weighed against the processing cost to determine if forage processing is warranted.

The processing of halophytes and salt tolerant plants may increase utilization of natural palatable halophytes or those less palatable with large biomass improve the nutritional value and palatability of forage plants with low nutritional value and palatability, provide balanced nutritional feed all year round.

Author details

Salah A. Attia-Ismail

Address all correspondence to: saai54@hotmail.com

Desert Research Center, Cairo, Egypt

References

[1] Aronson J. HALOPH: A Database of Salt Tolerant Plants of the World. Arizona, Tucson, AZ: Office of Arid Land Studies, University. 1985; p. 77

[2] Flowers TJ, Hajiheri MA, Clipson NJW. Halophytes. The Quarterly Review of Biology. 1986;**61**:313

[3] Squires VR, Ayoub AT. editors, Halophytes as a Resource for Livestock and for Rehabilitation of Degraded Land, Tasks in Vegetation Science Series. The Netherlands: Kluwer Academic Press; 1994

[4] Le Houerou HN. Forage halophytes and salt-tolerant fodder crops in the Mediterranean basin. In: Squires VR, Ayoub AT. editors. Halophytes as a Resource for Livestock and for Rehabilitation of Degraded Lands. 123 Kluwer Academic Publishers, Dordrecht; 1994

[5] Flowers TJ, Colmer TD. Salinity tolerance in halophytes. New Phytologist. 2008;**179**: 945-963

[6] Dudal R, Purnell MF. Land resources: Salt affected soils. Reclamation and Revegetation Research. 1986;**5**:19

[7] FAO. http://www.fao.org/docrep/x5871e/x5871e03.htm

[8] Shrestha DP, Farshad A. Mapping salinity hazard: An integrated application of remote sensing and modeling-based techniques. In "Remote sensing of soil salinization, Impact on land management". editor. Metternicht and Zinck. 6000 Broken Sound Parkway NW, Suite 300 Boca Raton, FL 33487-2742: CRC Press, Taylor & Francis Group; 2009

[9] Ramos J, Lopez MJ, Benlloch M. Effect of NaCl and KCl salts on the growth and solute accumulation of the halophyte *Atriplex nummularia*. Plant Soil. 2004;**259**:163-168

[10] Amouei, A. Effect of saline soil levels stresses on agronomic parameters and fodder value of the halophyte *Atriplex leucoclada* L. (Chenopodiaceae). African Journal of Agricultural Research. 2013;**8**(23):3007-3012

[11] Aronson JA. (1985). Economic halophytes—A global review. In: Wickens GE, Goodin JR, Field DV, editors. Plants for Arid lands. London: George Allen and Unwin; 1985; pp. 177-188

[12] Dagar JC. Characteristics of halophytic vegetation in India. In: Khan MA, Ungar IA, editors. Biology of Salt Tolerant Plants. Karachi: University of Karachi. 1995. pp. 55-76

[13] Ventura Y, Sagi M. Halophyte crop cultivation: The case for salicornia and sarcocornia. Environmental and Experimental Botany. 2013;**92**:144-153

[14] Attia-Ismail SA. Nutritional and feed value of halophytes and salt tolerant plants. In: El Shaer and Squires, editors. Halophytic and Salt Tolerant Feedstuffs: Impacts on Nutrition, Physiology and Reproduction of Livestock. 126: Vol. 106. 2015. CRC Press; New York.

[15] Attia-Ismail SA. Plant secondary metabolites of halophytes and salt tolerant plants. In: El Shaer and Squires, editors. Halophytic and Salt Tolerant Feedstuffs: Impacts on Nutrition, Physiology and Reproduction of Livestock. Vol. 127. CRC Press; 2015. p. 142

[16] Masters DG. Assessing the feeding value of halophytes. In: El Shaer and Squires, editors. Halophytic and Salt Tolerant Feedstuffs: Impacts on Nutrition, Physiology and Reproduction of Livestock. CRC Press; New York; 2015

[17] Le Houerou HN. Salt tolerant plants for the arid regions of the Mediterranean isoclimatic zone. In: Leith H, El-Masoom A, editors. Towards the Rational Use of High Salinity-Tolerant Plants. Vol. 1. Dordrecht, The Netherlands: Kluwer Academic Publications; 1993. p. 403

[18] Glenn EP, Brown J, Blumwald E. Salt tolerance and crop potential of halophytes. Critical Reviews in Plant Sciences. 1999;**18**:227-255

[19] Attia-Ismail SA. Role of minerals in halophyte feeding to ruminants. In: Prasad MNV, editor. Trace Elements: Nutritional Benefits, Environmental Contamination, and Health Implications. Wiley, Hoboken, New Jersey; 2008

[20] Attia-Ismail SA. Mineral balance in animals as affected by halophyte and salt tolerance plant feeding. In: El Shaer and Squires, editors. Halophytic and Salt Tolerant Feedstuffs: Impacts on Nutrition, Physiology and Reproduction of Livestock. Vol. 144. CRC Press; 2015. p. 157

[21] El-Shaer HM. Rangelands as feed resources in the Egyptian desert: Management and Improvement. Proceedings of the International Conference on Desert Development in the Arab Gulf Countries, State of Kuwait. 1996. pp. 23-26

[22] Raef O. Nutritive evaluation of natural ranges in the south eastern corner of Egypt. M.Sc. Faculty of Agriculture, Cairo University; 2012

[23] Albert R, Marianne P. Chemical compositions of halophytes from Neusiedler lake region in Australia. Oecologia (Berl.). 1977;**27**:157-170

[24] Gorham JLl, Hughes, Wyn Jones RG. Chemical composition of salt marsh plants from Ynys Mom (Anglesey): The concept of physiotypes. Plant, Cell and Environment. 1980; **3**:309-318

[25] Kearl LC. Nutrient requirements of ruminants in developing countries. Utah State University, Logan, Utah, USA: International Feedstuffs Institute. Utah Agricultural Experiment Station; 1982

[26] Agastian P, Kingsley SJ, Vivekanandan M. Effect of salinity on photosynthesis and biochemical characteristics in mulberry genotypes. Photosynthetica. 2000;**38**:287-290

[27] Wang J, Meng Y, Li B, Ma X, Lai Y, Si E, Yang K, Xu X, Shang X, Wang H, Wang D. Physiological and proteomic analyses of salt stress response in the halophyte halogeton glomeratus. Plant, Cell and Environment. 2014;**1**:15

[28] Hall JL, Flowers TJ. The effect of salt on protein synthesis in the halophyte Suaeda maritime. Panta (Berl.). 1973;**110**:361-368

[29] Kozlowski TT. Responses of woody plants to flooding and salinity. Tree Physiology Monograph No. 1. 1997

[30] Rani G. Changes in protein profile and amino acids in Cladophora vagabunda (Chlorophyceae) in response to salinity stress. Journal of Applied Phycology. 1997;**19**: 803-807

[31] Al-Khalasi SS. Osman Mahgoub, Isam Kadim T, Waleed Al-Marzooqi, Salim A. Al-Rawahi. Salt Tolerant Fodder for Omani Sheep (Effects of Salt Tolerant Sorghum on Performance, Carcass, Meat Quality and Health of Omani Sheep). A Monograph on Management of Salt-Affected Soils and Water for Sustainable Agriculture. 2010. pp. 67-81

[32] Benjamin RW, Oren E, Katz E, Becker K. The apparent digestibility of *Atriplex barclayana* and its effect on nitrogen balance in sheep. Journal of Animal Production Advances. 1992;**54**:259-264

[33] Masters DG. Establishing the metabolisable energy value of halophytic shrubs in vitro—problems and possibilities, confidential report. 2006

[34] Meneses R, Gabriel Varela, Hugo Flores. Evaluating the use of *Atriplex nummularia* hay on feed intake, growth, and carcass characteristics of creole kids. Chilean Journal of Agricultural Research. 2012;**72**(1):74-79

[35] Moinuddin M, Gulzar S, Aziz I, Alatar AA, Hegazy AK, Ajmal khan M. Evaluation of forage quality among coastal and inland grasses from Karachi. Pakistan Journal of Botany. 2012;**44**(2):573-577

[36] Thomas DT, Rintoul AJ, Masters DG. Sheep select combinations of high and low sodium chloride, energy and crude protein feed that improve their diet. Applied Animal Behaviour Science. 2007;**105**:140-153

[37] Shawkat, Safinaz.M. Khatab IM, Borhami BE, El-Shazly KA. Performance of growing goats fed halophytic pasture with different energy sources. Egyptian Journal of Nutrition and Feeds. 2001;**4**:251-326

[38] McEvoy JF, Jolly S. The nutritive value of rangelands plants of Southern and Western Australia: A review of the literature and a scoping study to demonstrate the inconsistencies between the current system of nutritive analysis and animal performance. AWI Project No: EC786. Productive Nutrition Pty Ltd. 2006

[39] Arab and Middle East Tables of Fed Composition. Utah Agricultural Experiment Station, Logan, Utah: International feedstuffs Institute; 1979

[40] Wilson AD. The intake and excretion of sodium by sheep fed on species of Ateriplex (saltbush) and Kochia (bluebush). Australian Journal of Agricultural Research. 1966;**17**: 155-163

[41] Attia-Ismail SA. Rumen physiology under high salt stress. In: El Shaer and Squires, editors. Halophytic and Salt Tolerant Feedstuffs: Impacts on Nutrition, Physiology and Reproduction of Livestock. Vol. 349. CRC Press; 2015. p. 357

[42] Degen AA, Squires VR. The rumen and its adaptation to salt tolerance. In: El Shaer and Squires, editors. Halophytic and Salt Tolerant Feedstuffs: Impacts on Nutrition, Physiology and Reproduction of Livestock. CRC Press; 2015

[43] Warner ACI, Stacy BD. Solutes in the rumen of the sheep. Quarterly Journal of Experimental Physiology. 1965;**50**:169-184

[44] Bergen WG. Rumen osmolality as a factor in feed intake control of sheep. Journal of Animal Science. 1972;**34**:1054-1060

[45] Kewan KZ. Studies on camel nutrition. Ph.D. Thesis. Egypt: Faculty of Agriculture - Alexandria University.

[46] Attia-Ismail SA, Elsayed HM, Asker AR, Zaki EA. Effect of different buffers on rumen kinetics of sheep fed halophyte plants. Journal of Environmental Science. 2009;**19**(1): 89-106

[47] Khamis HS. Nutritional studies on some agricultural by products and some natural pasture plants in arid and semi-arid areas using sheep and goats. Ph.D. Thesis, Faculty of Agriculture, Cairo University

[48] Gihad EA, El-Shaer HM. Utilization of halophytes by livestock on rangelands: Problems and prospects. In: Squires VR, Ayoub AT, editors, Halophytes as a Resource for Livestock and for Rehabilitation of Degraded Lands. Kluwer Academic Publishers; 1994. p. 77

[49] Attia-Ismail SA, Fayed AM, Fahmy AA. Some mineral, and nitrogen utilization of sheep fed salt plant and monensin. Egyptian Journal of Nutrition and Feeds. 2003;**6**:151-161

[50] Fahmy AA, Attia-Ismail SA, Afaf M Fayed. Effect of monensin on salt plant utilization and sheep performance. Egyptian Journal of Nutrition and Feeds. 2001;**4**:581-590

[51] Streeter MN, Hill GM, Wagner DG, Owens FN, Hibberd CA. Effect of bird resistant and nonbird resistant sorghum grains on amino acid digestion by beef heifers. Journal of Animal Science. 1993;**71**:1648-1656

[52] Fahmy AA. 2004. Unpublished data

[53] Sperry OE, Dollohite JW, Hoffman GG, Comp BJ. Texas poisonous plants to livestock. Texas. Agricultural Station Bulletin. 1964;**1028**

[54] Shalby AF, Etman MA, Yossef M, Habibi AM, Amer Kh F. A chemical investigation of Nitraria retusa Forsk. Desert Institute. Bulletin. Arab Republic of Egypt. 1977;**27**:199

[55] Van Soest PJ. Nutritional Ecology of the Ruminants. O and B Books. Corvallis, Oregon, USA: 1982. p. 374

[56] Jones TC, Hunt RC, King NW. Veterinary Pathology, 6th ed. Williams and Wilkins, Baltimore, Maryland; 1997

[57] El-Adawy TA. Nutritional composition and anti-nutritional factors of chickpeas (*Cicer arietinum* L.) undergoing different cooking methods and germination. Plant Foods for Human Nutrition (formerly Qualitas Plantarum). 2002;**57**(1):83-87

[58] Van Bruggen J, Veth P, Sebastiaan NL. Method and Device for Reducing the Amount of Anti-nutritional factors in a Mixture of Raw Material for Animal Feed. World Intellectual Property Organization (WO/1993/005664); 1993

[59] Attia-Ismail SA. Factors limiting and methods of improving nutritive and feeding values of halophytes in arid, semi arid and coastal areas. International Conference on Biosaline Agriculture & High Salinity Tolerance, Mugla, Turkey, 9-14 Jan, 2005, pp. 91-99

[60] Attia-Ismail SA, Fahmy AAM, Fouad RT. Improving nutritional value of some roughages with mufeed liquid supplement. Egyptian Journal of Animal Production. 1994; **31**:161-174

Chapter 5

An Insight into Current and Future Production of Forage Crops in Zimbabwe

Allan Sebata

Additional information is available at the end of the chapter

http://dx.doi.org/10.5772/intechopen.71997

Abstract

In Zimbabwe, most livestock are reared by smallholder farmers who live in marginal areas with low rainfall and hence poor forage production. As a consequence, livestock productivity is low and intermittent droughts result in animal mortalities. Forage crops have been widely promoted to provide feed resources to livestock, particularly during the dry season and in years of low precipitation. However, production of forage crops among the smallholder farmers remains low, especially in areas that receive low rainfall (<600 mm per annum). This chapter reviewed work on the production and promotion of forage crops in Zimbabwe in the past 50 years. The production and adaptation of different forage crops *viz.* improved grasses, herbaceous, and tree legumes to low and high (>800 mm per annum) rainfall areas is highlighted. Planting of improved grasses and herbaceous legumes in fallows and tree legumes as hedges and on contours hold the best promise in terms of improving the availability of forage crops for livestock feeding. Shortage of moisture remains the greatest constraint to increasing the area planted to forage crops. Therefore, the development of irrigation facilities needs to be encouraged to allow for the growing of forage crops.

Keywords: bana grass, ensiling, fallows, intercropping, reinforcement, tree legumes

1. Introduction

In Zimbabwe, the majority of the people live in rural areas with smallholder cropping and livestock production as the major sources of livelihoods. However, with the ever increasing population, the natural resources are constrained leading to low crop yields and livestock productivity. Livestock are reared on natural rangeland, which in most areas have been degraded leading to a decline in grass production. Thus, to improve livestock production, there is a need to promote forage crop production by smallholder farmers. Forage crops are plants grown for feeding livestock either directly as grazing pastures or as conserved hay and silage. They provide important nutrients (energy, protein, vitamins and minerals) to the livestock, particularly during the dry season when the natural grazing areas will be having poor quality forage.

The introduction of commercial smallholder dairy in Zimbabwe has seen an increase in the establishment of pasture grasses and legumes [1]. Some of the forage grasses grown include giant rhodes (*Chloris gayana*), Napier (*Pennisetum purpureum*), bana grass (*Pennisetum purpureum* cv. bana) and star grass (*Cynodon nlemfuensis*). Herbaceous legumes such as lablab (*Lablab purpureum*), fine-stem stylo (*Stylosanthes guianensis*), silverleaf desmodium (*Desmodium uncinatum*), siratro (*Macroptilium atropurpureum*) and velvet beans (*Mucuna pruriens*) have been established together with the grasses to improve dietary protein. In addition, multipurpose trees such as leucaena (*Leucaena leucocephala*) and sesbania (*Sesbania sesban*) are grown. However, the total area under forage crops remains very small (~4.2% of total arable land), mostly among the smallholder dairy farmers [2]. Chigariro [3] estimated that each individual smallholder dairy farmer was allocated between 0.4 (8% of arable landholding) and 0.8 ha (16% of arable landholding) for forage crop production in higher rainfall (>800 mm per annum) areas. However, an increasing number of small scale livestock keepers with irrigation facilities are now growing forage crops to feed their livestock during the dry season, which extends from June to November. This is meant to supplement the inadequate grass biomass and quality in natural rangelands particularly during drought years.

The aim of this chapter was to provide a review of the work done to date in Zimbabwe to promote forage crop production particularly in the smallholder farming sector. The chapter focused on reinforcing natural grassland with legumes, rehabilitating fallows with legumes, promotion of tree legumes as protein supplements, cropped forage grasses and legumes and forage crop conservation. The future prospects of forage crop production in the smallholder farming sector in Zimbabwe is also highlighted.

2. Reinforcing natural grassland with legumes

To improve animal production under natural rangeland conditions, strategies to increase grass production are required. One such strategy is grassland reinforcement with legumes, which increases both the quality and quantity of grazing. This has been found to improve the performance of individual animals and results in increased carrying capacity of the rangeland [4]. MacLaurin and Grant [4] reported cattle weight gains of over 60 percent per hectare (10,000 m^2) in rangelands reinforced with legumes as compared to the unimproved natural rangelands in Zimbabwe.

Selection of legume species adapted to the conditions in the high rainfall (>800 mm per year) areas of Zimbabwe was undertaken in the early 1960s [4]. Legumes found to be adaptable to these conditions included *Desmodium intortum* (greenleaf desmodium), *D. uncinatum* (silverleaf desmodium), *Macroptilium atropurpureum* (siratro), *Stylosanthes guianensis* var. intermedia (Oxley fine-stem stylo) and *Trifolium semipilosum* (Kenya white clover) [5]. *Stylosanthes guianensis* was found to establish and persist under adverse conditions [6]. It persisted under heavy grazing and increased its density through the establishment from shed seed [7]. Herbaceous legumes such as *Desmodium intortum, D. uncinatum* and *Macroptilium atropurpureum* have been found to persist under controlled grazing or cutting and produce yields of up to 10 tonnes dry matter per hectare per year on ploughed lands [8].

For the legumes to successfully establish, it is necessary to disturb the soil surface and to set back grass growth [6]. This can be done by burning-off of top hamper followed by disking the strips where the legumes are to be sown to improve emergence. The hard seeds of pasture legumes are scarified using hot water, dry heat or with concentrated sulphuric acid before planting [9].

Reinforcing *Hyparrhenia filipendula* grassland with *Stylosanthes guianensis* var. intermedia (Oxley stylo) and *Macroptilium atropurpureum* (siratro) increased dry matter yield by 58 and 49%, respectively, whereas crude protein increased by 64 (stylo) and 20% (siratro) (see **Table 1**) [10].

However, *S. guianensis* has been found to be intolerant to shading by the tall *H. filipendula* with long periods of rest [6]. Although, *M. atropurpureum* is more tolerant to shading, due to its twinning ability, it has poor dry matter yield when frequently grazed and will not persist under heavy grazing [4]. Clatworthy [11] recommended that grasslands reinforced with legumes be managed through rotational grazing with a grazing period of less than 2 weeks and rest period of at least 5 weeks. MacLaurin and Grant [4] reported steers grazing in grasslands reinforced with legumes as gaining 40 kg more than those on grassland only. However, weight gains of steers in grasslands reinforced with legumes were found to decline with decreasing rainfall [12].

A major constraint to grassland reinforcement with legumes has been the poor establishment of most species largely because of the low germination [4]. This has presumably led to the abandonment of reinforcement trials by research stations and farmers. However, it would be plausible to initiate new trials and broaden the legume species to be screened in view of the continued declining rangeland productivity. An alternative to natural grassland reinforcement would be to plant the legumes in fallow abandoned cultivated croplands. Tavirimirwa et al. [13] estimated that fallows constitute about 50% of the land, which was previously under

	Dry matter	**Crude protein**
H. filipendula only	5040	10.6
H. filipendula + stylo	7940	17.4
H. filipendula + siratro	7510	13.2

Source: Mufandeadza [10].

Table 1. Dry matter (kgha^{-1}) and crude protein (% dry matter) of *Hyparrhenia filipendula* grassland reinforced with *Stylosanthes guianensis* Var. Intermedia (stylo) and *Macroptilium atropurpureum* (siratro) in the high rainfall (>800 mm per annum) area of Zimbabwe.

cultivation in central Zimbabwe. The unavailability of legume species suited to the drier areas (<600 mm rainfall per year) is also a major challenge in grassland reinforcement with legumes. In addition, the nutrient requirements of the legumes in different soil types need to be determined [4].

Although herbaceous legumes produce high biomass and improve grassland dry matter yield in reinforcement trials, they offer poor foraging opportunities during the dry season because they shed leaves after frost and loose herbage as a result of trampling [8]. Hove et al. [14] found that herbaceous legumes such as *M. atropurpureum* and *S. guianensis* had been grown with limited success in the smallholder sector because of their lower productivity and failure to persist under low levels of management. This prompted the need to explore alternative fodder sources to improve availability of nutrient rich forage during the dry season.

3. Rehabilitating fallows with improved legumes

In most parts of Zimbabwe, large tracts of land are left to recover naturally following years of continuous cropping. Sowing improved legumes such as lablab or velvet bean (mucuna) has been found to accelerate the restoration to produce these fallows. Velvet beans grown on fallows yielded between 4700 and 11,300 kg/ha [15]. Both lablab and velvet bean crop can be grown, harvested and dried to make hay for feeding livestock or ploughed in as green mature to improve soil fertility. Planting forage legumes to restore fallow land improved maize grain yield by between 8 and 57%, which could be attributed to nitrogen fixation [15]. Therefore, the use of forage legumes in fallow restoration is beneficial in that large amounts of good quality forage is produced in addition to improved soil fertility. Muchadeyi [16] reported lablab crude protein content of about 12% of dry matter. The restoration of soil fertility is achieved through the fixation of atmospheric nitrogen and/or through an improvement in soil-physical properties [17]. In addition, lablab has been found to be drought tolerant with its deep and extensive root system contributing to soil organic matter content when decomposed, improving aeration and soil structure [15].

4. Use of tree forage legumes as protein supplements

In Zimbabwe, farmers have traditionally been feeding their livestock using leaves from native trees. However, in the last four decades, exotic (mostly from Central America) leguminous trees have been introduced because of their fast growth rates, acceptability to livestock and tolerance to frequent pruning and drought [18, 19]. In addition, these trees are long lived, require less maintenance and are palatable [20]. For example, *Acacia angustissima* is moderately palatable and digestible, *Calliandra calothyrsus* is palatable with low digestibility and *Leucaena leucocephala* is very palatable with high digestibility [21]. However, Hove [22] found *A. angustissima* and *C. calothyrsus* forage to be acceptable to livestock only after being dried. *Acacia angustissima, C. calothyrsus, Gliricidia sepium, L. leucocephala* and *Sesbania sesban* have been successfully introduced to smallholder dairy farmers in Zimbabwe [23]. Tree legumes are also referred to as multipurpose trees because they provide other products such as firewood and services such as soil erosion control [18].

The International Centre for Research in Agroforestry (ICRAF) in Zimbabwe tested and screened a number of leguminous tree forages suitability to provide forage to livestock particularly dairy cattle in the smallholder sector [24]. The biomass yield of five tree legumes in areas receiving high (>800 mm per annum) and low rainfall (<600 mm per annum) when cut once at the end of the rainy season is given in **Table 2**.

Leaf yields were higher in areas, which received high rainfall. *Acacia angustissima* had the highest leaf biomass and *S. sesban* the least (**Table 2**). Hove et al. [25] reported *C. calothyrsus* biomass yield of 2500–5600 kg/ha/year and *A. angustissima, L. leucocephala* and *G. sepium* yields of more than 3000 kg/ha/year. Dzowela et al. [24] recorded leaf dry matter yields of 5500, 3200 and 5800 kg/ha/year for 3-year old stands of *A. angustissima, C. calothyrsus* and *L. leucocephala* respectively. Matimati et al. [21] reported leaf yields of 400–3300, 800–5600 and 200–700 kg dry matter/ha/year for *A. angustissima, C. calothyrsus* and *L. leucocephala,* respectively, 1 year after establishment. They postulated that to get leaf dry matter yields of about 3500, 4000 and 1000 kg/ha/year for *A. angustissima, C. calothyrsus* and *L. leucocephala,* respectively required tree densities greater than 6000 trees/ha (equivalent to a tree spacing of 1.5 × 1.11 m). Tree legume leaf dry matter yield was found to increase with increasing frequency of moderate harvesting [26]. Muinga et al. [27] found supplementation of a basal diet of Napier grass with *L. leucocephala* to improve the digestibility of the Napier grass and to increase its intake by cross-bred dairy cows. Gusha et al. [20] reported crude protein content of 30.5, 26.5 and 22.7% of dry matter for *G. sepium, A. angustissima* and *C. calothyrsus*, respectively. Pigeon pea (*Cajanus cajan*) was successfully used to provide browse during the dry season at Grasslands Research Station, Zimbabwe, between 1951 and 1956 [8]. Steers fed pigeon pea forage gained an average of 0.22–0.34 kg per day over a 30- to 90-day period compared to those on fertilised grass pastures, which gained 0.09–0.18 kg per day for the same period [28].

Tree legumes can be grown as hedges along the field boundaries or on contours to limit soil erosion or on fallow land and as pure stands or intercropped with other crops [8]. In high

Species	High rainfall	Low rainfall
Acacia angustissima	3520–5530	3260–4850
Leucaena leucocephala	2850–5810	2060–5690
Gliricidia sepium	3040–5040	2430–6340
Calliandra calothyrsus	3210–3530	1240–3200
Sesbania sesban	1710–2980	920–1490

Source: Dzowela et al. [24].

Table 2. Leaf yields (kg/ha/year) of five leguminous fodder tree species in areas of high (>800 mm per annum) and low (<600 mm per annum) rainfall in Zimbabwe.

rainfall (>800 mm per annum) areas, most farmers intercropped the tree legumes with food (such as beans) and other fodder crops (*D. uncinatum, M. atropurpureum* and *S. guianensis*) [14]. Nyaata et al. [29] found that *C. calothyrsus* could be intercropped with Napier grass (*Pennisetum purpureum*) without reducing grass yields. To stimulate high biomass production, tree legumes can be grown for at least 1 year before being clipped at a height of 50 cm to initiate coppicing. The coppicing shoots can then be browsed by livestock, harvested and fed to the animals fresh or after drying [21].

Tree legumes are an important feed component in smallholder dairying. For instance, 1 kg of dried *C. calothyrsus* (24% crude protein and digestibility of 60% when fed fresh) was found to provide digestible protein equivalent to 1 kg of dairy meal (16% crude protein and 80% digestibility) [30]. *Calliandra calothyrsus* has also been found to be a good protein supplement to basal diets of Napier grass and crop residues [31]. Increases in milk production of 0.6–0.75 kg milk per kilogram of dried *C. calothyrsus* fed as supplement has been recorded [18].

Tree legumes could also play an important role in mitigating the effects of climate change because they are deep rooted, resistant to drought and are able to maintain high nutrient levels during the dry season [32]. In Zimbabwe, currently only a few farmers are exploiting the research findings of higher tree legume performances and the associated improvement in livestock production following their supplementation of basal diets of Napier grass or crop residues. Interestingly, high adoption rates have been reported in areas where tree legumes were introduced to smallholder farmers [14].

5. Cropped forage grasses

Livestock production can be improved by providing better nutrition through feeding good quality grasses [33]. The provision of good quality grasses maybe achieved through the use of improved grass species, such as Napier grass and its hybrids [34]. Napier or elephant grass (*Pennisetum purpureum*) and its hybrids are important forage crops particularly in areas that receive low rainfall because of their adaptations to drier conditions. They are robust perennial grasses that have been widely used as tropical forage producing higher dry matter yield than most tropical grasses [35]. Napier grass is a perennial grass that is indigenous to sub-Saharan Africa. It can withstand repeated defoliation as it regenerates quickly after cutting or grazing producing highly palatable foliage [36]. Napier grass and its hybrids have been widely adopted by smallholder farmers for feeding dairy cattle in most parts of Eastern, Central and Southern Africa [3, 37]. However, they have low protein concentration, requiring supplementation with legumes and protein concentrates [38]. The *Pennisetum* hybrids are produced by crossing Napier grass and pearl millet (*Pennisetum glaucum*) to improve the biomass yield and nutritive quality of the Napier grass. Gupta and Mhere [39] reported the interspecific hybrids as producing more tillers and leaves and having faster growth rates than their parents. The Napier × pearl millet hybrids (hereafter referred to as bana grass) are the most promising forage crops in the drier parts of Zimbabwe, offering high productivity, excellent quality and drought tolerance [39]. These hybrids are propagated vegetatively because they establish easily through developing shoots. Bana grass benefit from the desirable characteristics of pearl millet such as vigour, drought resistance and disease tolerance and those of Napier grass, which are good palatability and high dry matter yield.

The nutritive value of forage crops affects their utilisation by livestock, which in turn influences the production of the animals. Turano et al. [40] reported crude protein values of between 6.4 and 8.3% dry matter and *in vitro* dry matter digestibility of between 54.5 and 70% for Napier × pearl millet hybrids and Napier grass varieties. While silage crude protein content of 13.08% dry matter and *in vitro* dry matter digestibility of 66.93% were reported [41]. Bana grass can give dry matter yields of 10,000–12,000 kg per hectare and the best time to harvest is 6–7 weeks after onset of regrowth [42]. Nyambati et al. [43] reported mean annual dry matter yield of Napier grass varieties of 10,300 and 22,100 kg/ha/year over three and six harvests, respectively. In East Africa, values of 15,000–22,000 kg/ha/year were reported for Napier grass varieties [44]. Chigariro [3] reported increased livestock production with milk yields almost doubling following the use of bana/Napier forage crops in the high rainfall (>800 mm per annum) regions of Zimbabwe. Bana grass can be grown together with lablab and mucuna to improve crude protein content of animal diets.

Star grass (*Cynodon nlemfuensis*) is another important forage grass, which is adapted to low rainfall areas as it quickly produces foliage soon after the onset of the rains after the dry season [42]. Its crude protein content (5.54% of dry matter) at the end of the growing season was higher than that of grasses species found in natural grasslands (e.g. *Cynodon dactylon*—3.75% of dry matter) [13]. The crude protein content of 5.54% at the beginning of the dry season was high enough to mitigate against livestock weight losses making *C. nlemfuensis*, a suitable grass species to establish in fallow lands. In addition to the crude protein content, *C. nlemfuensis* is less fibrous than *C. dactylon* and other native grasses [13].

Forage sorghum (*Sorghum vulgare*) was bred specifically for feeding livestock. For instance, some varieties have sweet thin stems, are leafy and have good tillering ability. In addition, forage sorghum has comparable water soluble carbohydrates to maize (180–250 *vs.* 280–510 g/kg dry matter) [45, 46]. Water soluble carbohydrates are essential for successful ensilaging [47]. Forage sorghum silage has metabolisable energy of 9.5 MJ/kg dry matter compared to 10.2 MJ/kg dry matter for maize silage [48]. The sorghum silage was found to be of good fermentable quality, with a crude protein content of 12.0% of dry matter, when intercropped with lablab [46]. In addition, forage sorghum has been found to be adaptable to low rainfall producing high biomass yields [46].

6. Cropped herbaceous forage legumes

In Zimbabwe, a number of tropical herbaceous legumes have been introduced, screened and the most adapted selected as forage crops. Tropical herbaceous legumes have fast growth rates and perform well in unfertile sandy soils due to their ability to fix nitrogen. They are mainly used as green manure, in intercropping, crop rotation and as fodder for livestock. In addition, forage legumes have higher crude protein content than grasses [49]. For example,

Murungweni et al. [50] reported crude protein content of 17.3 and 20.3% of dry matter for dolichos bean (*Lablab purpureus*) (hereafter referred to as lablab) and velvet bean (*Mucuna pruriens*) (hereafter referred to as mucuna) hay, respectively, while most grasses have lower values (<13% of dry matter). Herbaceous forage legumes can be grown with food crops and then fed to livestock as supplements to crop residues and native pasture hay basal diets. Dry matter yields of 400–1058 and 345–1937 kg/ha/year for lablab and mucuna, respectively, can be attained in high rainfall areas (>800 mm per annum) [51]. In sub-Saharan Africa, mucuna and lablab are grown by smallholder farmers for feeding livestock [52]. In Zimbabwe, they have emerged as important forage or green-manure legumes for use in the smallholder crop–livestock systems [15, 51]. In addition, cowpea (*Vigna unguiculata*), which is traditionally grown as a food crop by smallholder farmers can be used to feed livestock to supplement native pasture hay diets.

Although research and promotion of forage legumes has been going on for more than half a century, smallholder farmers' adoption rates remain very low [52]. In view of the increasing demands for livestock products, strategies have to be put in place to improve adoption of forage legume production by smallholder farmers to increase animal production. One such strategy would be to identify farmers with irrigation facilities and willing to increase forage production. In addition, provision of information on the performance of different forage legumes under a wide range of environmental conditions can help farmers to make informed decisions. For instance, Murungweni et al. [50] reported a 15-fold increase in adoption rates of forage legume use by smallholder farmers in the high rainfall (>800 mm per annum) areas of Zimbabwe following on-farm trials. If this success can be replicated on a large scale, livestock production could increase substantially. Mucuna and lablab found to be the most adopted legumes for use in rotation with maize, restoration of fallow lands and for fodder production [34, 51]. The incentive for growing these forage legumes were enhanced soil fertility, which increased maize yield and provision of fodder for livestock. For example, maize yield doubled when the crop was grown a year after a legume [51]. The growing of legume forage crops also enabled smallholder farmers to replace commercial livestock feeds, reducing production costs. In addition to their good forage attributes, mucuna and lablab are also drought tolerant [51]. Furthermore, improved availability of seed could result in more farmers growing forage legumes.

Lablab is a high yielding forage crop with dry matter yields of about 10,900 kg/ha/year. It has a crude protein content of 14–19% of dry matter and low fibre content and high digestibility. Lablab can be fed as fresh foliage, hay or silage, although freshly harvested forage need to be wilted before feeding to avoid a bad flavour in the milk.

Mucuna is a tropical legume, which grows well in soils of low fertility because of its ability to fix nitrogen. It has a crude protein content of 11–23% of dry matter (dried beans contain 20–35% of dry matter crude protein) and has low fibre content.

Cowpeas are drought tolerant, grow well in sandy soils of low fertility and have high crude protein content (20–30% of dry matter). Three types of cowpeas have been developed *viz.* dual purpose, food type and forage type. Dual purpose cowpeas are semi-erect, produce average grain and forage yields, which can be used as food both for humans and for feeding livestock. The food type cowpeas varieties are erect with high pod yield and low vegetative biomass, while the forage type is spreading/prostrate with low pod yield but high biomass yield. Cowpeas are highly palatable to livestock with high intake and digestibility due to the high crude protein, vitamins and mineral content. They can be fed to livestock as fresh forage, hay or silage.

7. Forage crop conservation

Forage crops can be conserved as hay or silage for feeding livestock during the dry season when the grass in the natural rangeland would have declined in nutritive quality. In dairy farming, silage is the more preferred method of conserving forage crops. The aim of making silage is to preserve the energy and protein content of the forage for feeding during the dry season. The forage crops are harvested at an early growth stage when their nutrient quality is still high. Silage making is divided into three stages *viz.* forage harvesting and transporting, chopping and compaction and air sealing. Silage is made by putting freshly cut forage into a sealed place such as a pit covered with plastic or a plastic bag called a silo. The forage material is chopped into small pieces and compressed in the silo to remove air to create anaerobic conditions. The silo is then completely sealed to prevent air entry. Under these anaerobic conditions, lactic acid bacteria convert some of the sugars in the forage crop into lactic acid. The nutrient value of good silage should be comparable to that of the forage used to make it. A good silage has a pH of less than 5.0, the percent of total nitrogen which is ammonia (NH_3N:N) of less than 15%, lactic acid which is 50% or more of the total organic acids and butyric acid content of not greater than 0.5% of the total dry matter.

When selecting material for making silage, the following need to be considered; plant material nutritive value (energy, protein, vitamins and minerals), easy of ensiling and most suited crop for the area (rainfall, soil type, temperature, day length). The amount of silage to be made will depend on the number of animals to be fed, length of feeding period, proportion of silage in total ration and equipment available on the farm.

Since maize is a staple food for most households in Zimbabwe, other grass plants such as bana grass can be used to make silage. To get the best quality silage from bana grass, harvesting should be before it develops internodes (6–8 weeks after onset of resprouting or shoot development). Dry matter content should be between 30 and 40% to get good compaction. Only forage that will be ensiled should be cut on each occasion to minimise wilting loses. Ensiling should be done in as short a time as possible preferably within a day to preserve forage quality. The plant material should be fine and evenly chopped to maintain forage quality. For instance, it can be chopped into 2 cm pieces using a motorised chopper or into 4–5 cm pieces manually using axes. The material should be compacted tightly to remove all the air out of the silo. After compaction, the silo should be sealed tightly to prevent air entrance. All the sides of the silo should be covered with polyethylene and soil put on top of the cover to keep it air tight and avoid soil and water entrance.

Cereal crops such as maize, forage sorghum and bana grass can be ensiled with legumes to improve silage quality. Ensiling tree legumes mixed with maize or Napier grass improves the energy and protein content of the silage [52]. Cereal silages are rich in energy but low in crude protein while the converse is true for legume silages [53]. Ensiling maize with legumes increases the crude protein content of the silage. For instance, crude protein content increased from 7.7% of dry matter for maize silage only to 9.3–15.3% of dry matter with a legume incorporated [54]. Maasdorp and Titterton [55] ensiled the leaf material of four forage tree legumes with maize, on a fresh mass ratio of 50:50 (w/w) and the crude protein content increased to 14.0, 15.5, 17.2 and 18.7% of dry matter for maize-*Calliandra calothyrsus*, maize-*Gliricidia sepium*, maize-*Leucaena leucocephala* and maize-*Acacia boliviana* silages, respectively. The fermentation characteristics of *A. boliviana*-maize and *L. leucocephala*-maize silages are shown in **Table 3**.

The crude protein content of ABM and LLM silage were above the minimum requirement for growth (11.3% of dry matter) in ruminant animals [57]. Titterton et al. [52] reported higher crude protein content values of 20.87 and 17.60% of dry matter for ABM and LLM respectively. The metabolisable energy content of the silages was above the minimum acceptable level of 8 MJ/kg dry matter required for maintenance [58]. Ensiling increased the modified acid detergent fibre but reduced the neutral detergent fibre content of the silages.

Although ensiling cereal crops with tree legumes improves the quality of the silage, it has not been widely promoted among the smallholder farmers. For example, Hove et al. [14] reported a few farmers (7%) as ensiling tree legume leaves mixed with maize in high rainfall (>800 mm per annum) areas of Zimbabwe.

Forage crops can also be harvested during the wet season and preserved as hay for feeding animals during the dry season. To make hay, the forage crops are harvested at 50% flowering stage and air dried for a day or two to reduce moisture content. Legumes such as cowpea, lablab and mucuna can be air dried and rolled into bundles and stored in shades with good ventilation.

Constituent	**Before ensiling**			**After ensiling**		
	MS	**ABM**	**LLM**	**MS**	**ABM**	**LLM**
Crude protein	7.6	15.6	14.1	6.9	14.8	13.2
Modified acid detergent fibre	28.0	28.5	28.8	31.1	33.7	34.8
Neutral detergent fibre	51.6	52.1	52.3	49.9	49.1	47.7
Acid detergent fibre	36.3	36.1	36.9	35.0	34.6	35.3
Ash content	6.0	6.2	6.2	5.8	5.6	5.7
Metabolisable Energy/MJ/kg dry matter	11	10	10	10	10	9

Source: Phiri et al. [56].

Table 3. Chemical composition (% of dry matter) of maize silage (MS), *Acacia boliviana*-maize (ABM) and *Leucaena leucocephala*-maize (LLM) mixed silages before and after ensiling.

8. Future prospects of forage crop production

In Zimbabwe, currently less than 3% of smallholder farmers are growing forage crops [2]. This inevitably means low livestock productivity due to feed shortages, particularly during the dry season. However, farmers with smallholder dairy projects in areas receiving medium rainfall (600–800 mm) grew between 0.1 and 1 ha of fodder crops [1]. Forage and browse legumes grown were *M. pruriens, L. purpureus, V. unguiculata, M. atropurpureum, S. guianensis, L. leucocephala, A. angustissima, C. calothyrsus, C. cajan* and *S. sesban* while the most common cultivated grasses were *Pennisetum* spps and *C. nlemfuensis*. This suggests that large scale promotion could see more farmers growing forage crops, which could improve feed availability to livestock. However, most parts of Zimbabwe receive low and erratic rainfall and farmers have no irrigation facilities, which constrain production of forage crops. Thus, to increase forage crop production in areas receiving low rainfall farmers need to invest in irrigation facilities. For instance, dry matter yields of Napier grass varieties and pearl millet × Napier grass hybrids ranged from 10,300 to 32,100 kg/ha/year without irrigation and 19,600–55,800 kg/ha/year with irrigation, evidence that moisture stress can reduce biomass yield of rain fed forage crops [40]. Furthermore, increases in forage crop production can be achieved through farmer training, improved supply of seed and other inputs such as rhizobial inoculants [42].

Author details

Allan Sebata

Address all correspondence to: allan.sebata@nust.ac.zw

Department of Forest Resources and Wildlife Management, Faculty of Applied Sciences, National University of Science and Technology, Zimbabwe

References

[1] Mapiye C, Foti R, Chikumba N, Poshiwa X, Mwale M, Chivuraise C, Mupangwa JF. Constraints to adoption of forage and browse legumes by smallholder dairy farmers in Zimbabwe. Livestock Research for Rural Development. 2006;**18**:1-6

[2] Hanyani-Mlambo BT, Sibanda S, Rukuni M. Farming Systems in Zimbabwe's Smallholder Agricultural Sector: The Case of Gokwe South District. Department of Agricultural Economics and Extension, University of Zimbabwe, Zimbabwe; 2000

[3] Chigariro B. The farming Systems of the Zimbabwean District of Wedza and the role and adoption of forage legumes in small-scale crop–livestock enterprises. In: Whitbread AM, Pengelly BC, editors. Tropical Legumes for Sustainable Farming Systems in Southern Africa and Australia. Canberra, Australia: Australian Centre for International Agricultural Research (ACIAR) Proceedings No. 115; 2004 180 pp

[4] MacLaurin AR, Grant PJ. Research experience of reinforcing veld with legumes in Zimbabwe. In: Kategile JA, Said AN, Dzowela BH, editors. Animal Feed Resources for Small-Scale Livestock Producers - Proceedings of the Second Pastures Network for Eastern and Southern Africa (PANESA) Workshop, Held at the International Laboratory for Research on Animal Diseases, Kabete, Nairobi, Kenya, 11-15 November 1985. Kenya: International Development Research Centre; 1985

[5] Barnes DL. The preliminary evaluation of grasses and legumes for use in the high rainfall sandveld of Rhodesia. Proceedings of the Grassland Society of Southern Africa. 1966;**1**:69-74

[6] Clatworthy JN, Thomas PI. Establishment of *Stylosanthes guianensis* in Marandellas sandveld. Proceedings of the Grassland Society of Southern Africa. 1972;**7**:76-83

[7] Clatworthy JN. Screening of legumes for veld reinforcement. Proceedings of the Grassland Society of Southern Africa. 1980;**15**:111-116

[8] Clatworthy JN, Maclaurin AR, Avila M. The possible role of forage legumes in communal area farming systems in Zimbabwe. In: Kategile JA, Said AN, Dzowela BH, editors. Animal Feed Resources for Small-Scale Livestock Producers - Proceedings of the Second Pastures Network for Eastern and Southern Africa (PANESA) Workshop, Held at the International Laboratory for Research on Animal Diseases, Kabete, Nairobi, Kenya, 11-15 November 1985. Kenya: International Development Research Centre; 1985

[9] Grant PJ, Clatworthy JN. How to establish pasture legumes. Rhodesian Farmer. 1978; **48**:13-19

[10] Mufandeadza OT. Effect of Frequency and Height of Cutting on the Yield, Composition and Quality of the Herbage of Grass Grown Alone or in Mixture with Legumes. Annual Report 1976/77. Zimbabwe: Division of Livestock and Pastures, Department of Research and Specialist Services; 1978. pp. 173-176

[11] Clatworthy JN. Pasture research in Zimbabwe: 1964-1984. In: Kategile JA, editor. Pasture Improvement Research in Eastern and Southern Africa. Ottawa, Canada: IDRC; 1985. pp. 25-58

[12] Kelly RD. Tiffin, JdeW. Effect on Animal Production of Veld Reinforced with Siratro or Fine-Stem Stylo in a Medium-Rainfall Area. Annual Report 1981/82. Zimbabwe: Division of Livestock and Pastures, Research and Specialist Services; 1984. pp. 163-164

[13] Tavirimirwa B, Manzungu E, Ncube S. The evaluation of dry season nutritive value of dominant and improved grasses in fallows in Chivi district, Zimbabwe. Online Journal of Animal and Feed Research. 2012;**2**:470-474

[14] Hove L, Franzel S, Moyo PS. Farmer experiences in the production and utilisation of fodder trees in Zimbabwe: Constraints and opportunities for increased adoption. Tropical Grasslands. 2003;**37**:279-283

[15] Jiri O, Maasdorp BV, Temba E, Whitbread AM. The role of dual purpose and forage legumes to improve soil fertility and provide high quality forage: Options in a maize-based

system in Zimbabwe. In: Whitbread AM, Pengelly BC, editors. Tropical Legumes for Sustainable Farming Systems in Southern Africa and Australia. Canberra, Australia: Australian Centre for International Agricultural Research (ACIAR) Proceedings No. 115; 2004. 180 pp

[16] Muchadeyi R. Herbage yields, chemical composition and *in vitro* digestibility of dual purpose legumes intercropped with maize for dry season fodder supplementation. MSc thesis, University of Zimbabwe. 1998

[17] Wilson JR. Advances in nitrogen cycling in agricultural ecosystems. In: Proceedings of the Symposium on Advances in Nitrogen Cycling in Agricultural Ecosystems, Brisbane, Australia, 11-15 May 1987. Wallingford, UK: CAB International; 1988. p. 333

[18] Franzel S, Carsan S, Lukuyu B, Sinja J, Wambugu C. Fodder trees for improving livestock productivity and smallholder livelihoods in Africa. Current Opinion in Environment Sustainability. 2014;**6**:98-103

[19] Hess HD, Tiemann TT, Noto F, Franzel S, Lascano C, Kreuzer M. The effects of cultivation site on forage quality of *Calliandra calothyrsus* var. Patulul. Agroforestry Systems. 2006;**68**:209-220

[20] Gusha J, Ngongoni NT, Halimani TE. Nutritional composition and effective degradability of four forage trees grown for protein supplementation. Online Journal of Animal and Feed Research. 2013;**3**:170-175

[21] Matimati I, Maasdorp BV, Hove L. On-farm productivity of *Acacia angustissima, Calliandra calothyrsus* and *Leucaena leucocephala* in a subhumid area in Zimbabwe. African Journal of Range and Forage Science. 2009;**26**:75-80

[22] Hove L. Growing Trees for Profitable Milk Production. World Agroforestry Centre: Nairobi; 2004

[23] Chakeredza S, Hove L, Akinnifesi FK, Franzel S, Ajayi O, Sileshi G. Managing fodder trees as a solution to human–livestock food conflicts and their contribution to income generation for smallholder farmers in southern Africa. National Resource Forum. 2007;**34**:286-296

[24] Dzowela BH, Hove L, Mafongoya PL, Maasdorp BV. Recent work on the establishment, production and utilization of multipurpose trees as a feed resource in Zimbabwe. Animal Feed Science and Technology. 1997;**69**:1-15

[25] Hove L, Ndlovu L, Sibanda S. The effects of drying temperature on chemical composition and nutritive value of some tropical fodder shrubs. Agroforestry Systems. 2003;**59**: 231-241

[26] Matimati I. Determining Stage of Cutting for *A. boliviana* (Syn. *A. angustissima*), *C. calothyrsus* and *L. leucocephala* for Optimising Leaf Production. BSc Agric (Hons) Thesis. Zimbabwe: University of Zimbabwe; 1999

[27] Muinga RW, Topps JH, Rooke JA, Thorpe W. The effect of supplementation with *Leucaena leucocephala* and maize bran on voluntary food intake, digestibility, live weight

and milk yield of *Bos indicus* × *Bos taurus* dairy cows and rumen fermentation in steers offered *Pennisetum purpureum ad libitum* in the semi-humid tropics. Animal Science. 1995;**60**:13-23

[28] Mills PFL. Pigeon pea as a pasture legume in the high rainfall sandveld regions of southern Rhodesia. Rhodesia Agricultural Journal. 1961;**58**:171-172

[29] Nyaata OZ, O'Neil MK, Roothaert RL. Comparison of *Leucaena leucocephala* with *Calliandra calothyrsus* in Napier (*Pennisetum purpureum*) fodder banks. In: Shelton HM, Gutteridge RC, Mullen BF, Bray RA, editors. Leucaena-Adaptation Quality and Farming Systems. Vol. 86. ACIAR Proceedings. Proceedings of a Workshop Held in Hanoi, Vietnam, 9-14 February; 1998. pp. 257-260

[30] Roothaert R, Franzel S, Muriuki K. On-farm evaluation of fodder trees and shrubs preferred by farmers in central Kenya. Experimental Agriculture. 2003;**39**:423-440

[31] Paterson RT, Karanja GM, Roothaert R, Nyaata Z, Kariuki IW. A review of tree fodder production and utilization within smallholder agroforestry systems in Kenya. Agroforestry Systems. 1998;**41**:181-199

[32] Wambugu C, Place F, Franzel S. Research, development and scaling up the adoption of fodder shrub innovations in East Africa. International Journal of Agricultural Sustainability. 2011;**9**:100-109

[33] Humphreys LR. Tropical Pasture Utilisation. Cambridge, UK: Cambridge University Press; 1991. p. 206

[34] Pengelly BC, Whitbread A, Mazaiwana PR, Mukombe N. Better use of research resources to deliver adoption and benefits to farmers. In: Whitbread AM, Pengelly BC, editors. Tropical Legumes for Sustainable Farming Systems in Southern Africa and Australia. Canberra, Australia: Australian Centre for International Agricultural Research (ACIAR) Proceedings No. 115; 2004. 80 pp

[35] Hanna WW, Chaparro CJ, Mathews BW, Burns JC, Sollenberger LE, Carpenter JR. Perennial Pennisetums. In: Moser LE, Burson BL, Sollenberger LE, editors. Warm-Season (C4) Grasses. Madison, WI, USA: American Society of Agronomy Monograph Series No. 45; 2004. pp. 503-535

[36] Lowe AJ, Thorpe W, Teale A, Hanson J. Characterisation of germplasm accessions of Napier grass (*Pennisetum purpureum* and *P. purpureum* x *P. glaucum* hybrids) and comparison with farm clones using RAPD. Genetic Resources and Crop Evolution. 2003;**50**:121-132

[37] Valk YS. Review Report of the DEAF Surveys during 1989 NDDP/M41/ 200. Nairobi, Kenya: Ministry of Livestock Development; 1990

[38] Nyambati EM, Sollenberger LE, Kunkle WE. Feed intake and lactation performance of dairy cows offered Napier grass supplemented with legume hay. Livestock Production Science. 2003;**83**:179-189

[39] Gupta SC, Mhere O. Identification of superior pearl millet by Napier hybrids and napiers in Zimbabwe. African Crop Science Journal. 1997;**5**:229-237

[40] Turano B, Tiwari UP, Jha R. Growth and nutritional evaluation of napier grass hybrids as forage for ruminants. Tropical Grasslands. 2016;**4**:168-178

[41] Manyawu GJ, Sibanda S, Mutisi C, Chakoma IC, Ndiweni PNB. The effect of pre-wilting and incorporation of maize meal on the fermentation of bana grass silage. Asian-Australian Journal of Animal Science. 2003;**16**:843-851

[42] Mapiye C, Mupangwa JF, Mugabe PH, Chikumba N, Poshiwa X, Foti R. A review of improved forage grasses in Zimbabwe. Tropical and Subtropical Agroecosystems. 2006; **6**:125-131

[43] Nyambati EM, Muyekho FN, Onginjo E, Lusweti M. Production, characterization and nutritional quality of Napier grass [*Pennisetum purpureum* (Schum.)]. African Journal of Plant Science. 2010;**4**:496-502

[44] Muia JMK, Taminga S, Mbugua PN, Kariuki JN. Effect of supplementing Napier grass (*Pennisetum purpureum*) with poultry litter and sunflower meal based concentrates on feed intake and rumen fermentation in Friesian steers. Animal Feed Science and Technology. 2001;**92**:113-126

[45] McDonald P, Henderson AR, Hermon SJE. The Biochemistry of Silage. New York: A Wiley-Interscience Publication; 1991

[46] Mhere O, Maasdorp BV, Titterton M. In: Smith T, editor. Proceeding of Workshop on Livestock Production Research in the Semi-Arid Tropics. Matopos, Zimbabwe, United Kingdom: Department for International Development (DFID); 1999

[47] Titterton M, Bareeba FB. Grass and legume silages in the tropics. In: Silage Making in the Tropics with Particular Emphasis on Smallholders. Proceedings of the FAO Electronic Conference on Tropical Silage ('T Mannetje, L. Eds). September–December 1999. 2000

[48] Titterton M. The feasibility of using forage sorghum in drought years for silage in Zimbabwe. In: Kirk R, editor. Proceedings of Annual General Meeting on Forage Production for Dairying in Zimbabwe. Grasslands Society of Zimbabwe. p. 1997

[49] Nyoka R, Chikumba N, Chakoma I, Mazaiwana P, Mukombe N, Magwenzi N. Evaluation and screening of forage legumes for sustainable integration into crop–livestock farming systems of Wedza district. In: Whitbread AM, Pengelly BC, editors. Tropical Legumes for Sustainable Farming Systems in Southern Africa and Australia. Canberra, Australia: Australian Centre for International Agricultural Research (ACIAR) Proceedings No. 115; 2004. 180 pp

[50] Murungweni C, Mabuku O, Manyawu GJ. Mucuna, lablab and paprika calyx as substitutes for commercial protein sources used in dairy and pen-fattening diets by smallholder farmers of Zimbabwe. In: Whitbread AM, Pengelly BC, editors. Tropical Legumes

for Sustainable Farming Systems in Southern Africa and Australia. Canberra, Australia: Australian Centre for International Agricultural Research (ACIAR) Proceedings No. 115; 2004. 180 pp

[51] Manyawu GJ, Jiri O, Chigariro B, Mazaiwana PR, Nyoka R, Chikazunga D. A socioeconomic evaluation of an ACIAR-funded project on the use of forage legumes in integrated crop–livestock systems on smallholder farms in Zimbabwe. In: Whitbread AM, Pengelly BC, editors. Tropical Legumes for Sustainable Farming Systems in Southern Africa and Australia. Canberra, Australia: Australian Centre for International Agricultural Research (ACIAR) Proceedings No. 115; 2004. 180 pp

[52] Titterton M, Maasdorp BV, Mhere O, Nyoni L. Forage production and conservation for dry season feeding of dairy cows in semi-arid region of Zimbabwe. In: Smith T, Mhunduru J, Richards JI, editors. Proceedings of the Second Workshop on Livestock Production Programme Projects in Zimbabwe, (February, 1999). Chatham, UK: NRIL; 2000. pp. 23-34

[53] Catchpole VR, Henzell EF. Silage and silage making from tropical herbage species. Herbage Abstracts. 1971;**41**:213-221

[54] Titterton M, Maasdorp BV. Nutritional improvement of maize silage for dairying: Mixed crops silages from sole and intercropped legumes and a long season variety of maize. 2. Ensilage. Animal Feed Science and Technology. 1997;**69**:263-270

[55] Maasdorp BV, Titterton M. The use of planted trees for fodder. In: Smith T, editor. Proceeding of Workshop on Livestock Production Research in the Semi-Arid Tropics. Matopos Zimbabwe: Department for International Development (DFID); 1999

[56] Phiri MS, Ngongoni NT, Maasdorp BV, Titterton M, Mupangwa JF, Sebata A. Ensiling characteristics and feeding value of silage made from browse tree legume-maize mixtures. Tropical and Subtropical Agroecosystems. 2007;**7**:149-156

[57] ARC (Agricultural Research Council). The Nutrient Requirements of Livestock. Commonwealth Agricultural Bureaux. England: Slough; 1984

[58] Ekern A, Vik-Mo I. Conserved forage as feeds for dairy cows. In: Broster WH, Swan H, editors. Feeding Strategy for the High Yielding Cow. London: Grenada Publications; 1979

Section 2

Aspects of Forage Cultivation

Chapter 6

Phosphorus in Forage Production

Ademar Pereira Serra, Marlene Estevão Marchetti,
Elisângela Dupas, Carla Eloize Carducci,
Eulene Francisco da Silva,
Elaine Reis Pinheiro Lourente,
Alessandra Mayumi Tokura Alovisi and
Simone Candido Ensinas

Additional information is available at the end of the chapter

http://dx.doi.org/10.5772/intechopen.70202

Abstract

The aim in developing this work was to summarize information about phosphorus (P) limitation and dynamic in tropical soils for forage grasses production. The major idea is direct information about limited factors affecting P availability, dynamic of P fractionation, P pools, P forms, P use efficiency, and the 4R's Nutrient Stewardship' for P-fertilizer in forage grasses. Organizing these sub-headings in a chapter can result in interesting of how P behaves under tropical soils, in order to take decision to manage P-fertilizer to accomplish forage grasses production with social, economic, and environmental benefits. As the most limiting nutrient in tropical soils, P-fertilizer in forage grasses can be more effective if the best management practices are followed. In order to avoid excess P-fertilizer application in soil or P-fertilizer response with low efficiency, it is important to understand the P dynamic and the factors associated with P adsorption in soil. Even with low amount of P requested to forages species, the P available in soil is quite low, and this knowledge is primordial to direct P-fertilizer. Tropical soils are quite limited in P content, due to the natural formation with parental material poor in P content and highly weathering condition. Thus, in order to improve phosphorus use efficiency, the 4R's must be followed to improve P use efficiency (PUE). It is not easy to improve PUE in highly weathering soil with high buffering capacity; however, all the combination of best management practices for P-fertilizer application can result in better use efficiency. Based on the scarcity of natural P-sources in the whole world, the use of alternative P-sources should be incentivized, and more researches about this issue are need for better understanding.

Keywords: soil fertility, P-use efficiency, *Brachiaria* spp., *Panicum maximum*, *Stylosanthes* spp.

1. Introduction

The essentiality of phosphorus (P) for plants, grazing animal, and human being is quite evident [1–3]. The matter is the amount of P reservoir lasting to sustain the growing population in earth. Phosphorus reservoir in the world is decreasing, and the consumption of P-fertilizer is increasing worldwide [4], which is a challenge for farmers cropping in tropical regions under highly weathering soil with high P-sorption capacity and low P availability [5].

Phosphorus is considered the major limited factor in forage grasses production followed by nitrogen in tropical climate. Phosphorus availability in tropical soils is quite low due to much factors associated with P pools. Even with the resistance of *Brachiaria* spp. in support of low amount of P content in soil [6], the P constraint can decrease the biomass production and persistence of this genus in grassland. *Brachiaria* spp. are widely cultivated in Brazil, and most part of tropical climate, which support the beef cattle production under low cost for farmers due to the capacity of forage regrowth and permanence in the field even in winter season [7]. The utilization of P-fertilizer is not widely practiced in extensive pastures, resulting in pasture degradation, which can be classified as stocking rate capacity <0.4, 0.4–0.8, and 0.8–1.5 AU ha^{-1}, respectively, highly degraded, moderated degraded, and soft degraded [8]. Brazil has millions of hectares with degraded pastures, resulting in small stocking rate and unsustainable production system.

Brazilian Cerrado is considered one of the last frontiers for agricultural land, and remaining degraded areas with low food production are no longer be allowed due to limitation of agricultural areas in the world and increasing demand for food in a growing population. The integrated crop-livestock system has been taken place of degraded pastures in Brazilian Cerrado, resulting in improvement of stocking rate due to better soil fertility for forages cultivated right after grain crops. In this case, forages are cultivated without expend in P-fertilizer because of residual P from grain crops cultivation is enough. In integrated crop-livestock system, there are possibilities to recover pasture degradation with grain crops and remain the pasture for more than 1 year in the same area; thus, the forage is introduced into a crop rotation system [9]. The benefits of forages grasses for soil physical and chemical properties are very considerable, as the improvement of soil organic matter (SOM), straws on soil surface, soil aggregate stability, including nutrient recycling and other improvement in soil properties [10, 11].

Despite the integration production systems, the most area cultivated with pastures under extensive livestock management are not well supplied with P-fertilizer. The potential to improve biomass production through P-fertilizer is quite evident due to P deficiency in highly weathering soil of tropical climate. The replacement of P extracted by grazing animals is not accomplished as it should be [12]. Phosphorus fertilizer in forages improves its capacity of tillering associated with faster and higher biomass production. The answer of forage depends on species cultivated; the most common species in tropical region are *Brachiaria* spp., *Panicum* spp., and *Stylosanthes* spp., which majority cultivars were developed by Brazilian Agricultural Research Corporation (Embrapa) through its forage breeding program [13].

Phosphorus fertilizer alone cannot recover a pasture degraded, because we must keep in mind that soil compaction, soil pH, and availability of other essential nutrients can affect the phosphorus use efficiency (PUE), and consequently, the P-fertilizer management need to be well planned based on best management practice. The 4R's (right source, right rate, right place, and right time) nutrient stewardship is a very useful tool for fertilizer management [14], because it takes accounts of the social, environmental, and economic benefits that the fertilizer practice can promote. In order to improve PUE through the application of 4R's nutrient stewardship concepts, it is important to adjust in site-specific location.

In order to improve dry matter production, soil P content must be corrected and maintained for plant nutritional demands. The aim of this chapter is to summarize information about phosphorus (P) limitations, and that it is dynamic in tropical soil for forage grasses.

2. Dynamic of phosphorus in tropical soils

2.1. Phosphorus pool in soils

Most of tropical and subtropical regions have highly weathered soils, resulting in bases leaching and formation of Fe and Al oxides (the term includes oxides, hydroxides, and oxide-hydroxides). Iron and Al oxides contribute to increase positive charges in soil and consequently adsorption of anion, as the case of phosphate ($H_2PO_4^-$ and HPO_4^-) [15]. Phosphorus is in soil under different forms, which were proposed different methods of fractionation usually divided into two pools, organic P (Po) and inorganic P (Pi).

3. Organic phosphorus

The availability of Po is directly related to soil carbon dynamic [16]. Soil organic matter mineralization contributes to 60–80% of the total P available in soil, and this speed of P release rate depends mostly on C/N ratio of straw, thus forage grasses with higher C/N ratio contribute to accumulate straw above and below ground, resulting in increasing SOM. Besides CO_2 sequestration, root decomposition is the majority route of carbon entrance in soil, which is 21.2% of soil carbon, especially in relation to grass that has higher root density associated with higher renew rate of root [17]; thus, forages can contribute to great increment of SOM in soil profile.

Phosphorus organic sources can increase the P content in solution through the mineralization of straws and consequently releasing P in soil solution. On the other hand, temporary immobilization of Pi can occur in soil through the incorporation of Pi in microbial biomass (source of energy), increased by addition of carbon source as forage grass straws with limited concentration of P to offer for microbial population growth. Therefore, to mineralize these straws is necessary initial immobilization of Pi for while related to decreasing of carbon source, resulting in decreasing C/P ration of straws to values close to microbial biomass C/P ratio [15]. Consequently, the Pi in solution tends to increase with the stabilization of C/P in straws and microbial biomass.

Organic P and carbon are quite related to structured soil components (granulometric fractions, mineralogy, aggregate), which are stored in macroaggregate compounds of microaggregates formed by fresh fraction occlude of SOM [16, 18]. However, Po content can decrease faster in soils from tropical region than temperate regions due to higher rate of SOM mineralization, which are associated with higher temperature and rainfall [19, 20]. In tropical climate under forage grasses, 60% of Po in soil can be converted in P enzymatic against 36% in temperate climate [16].

The soil tillage and use as well as native vegetation can alter the P forms in the soil colloids, majority the organic forms, due to direct relation to soil biological activity. When a forestry is converted to a grassland and grain crops area, there are direct alteration in nutrient cycle. The transformation of Po into Pi is through the phosphatase enzyme. The phosphatase enzymes are excreted by plant roots and microorganisms in soil, as well as the deposition of straws on soil surface that increases the content of SOM [18, 19]. Organic P comprehended more than 50% of the total P in soil in agricultural systems [2], and its content depends on soil tillage and use. The hydrolase of Po and release of orthophosphate by enzymatic activity of phosphatase are affected by many factors, including the nature of Po, capacity of Po to interact with soil matrix, presence of microorganism that facilitated the mineralization, soil mineralogy, and physical–chemical soil properties [21].

4. Inorganic phosphorus

Phosphorus cycle, differently from C, N and S cycle, involves equilibrium reaction among the organic and inorganic constituents of soil. Besides P be considered the macronutrient less required in quantity by forage grass and other plant species, it is the most limited nutrient by plant growth in tropical climate due to higher P-sorption with Al/Fe oxides colloids in highly weathered soils [22]. The magnitude of P-sorption is related to the type and quantity of adsorption sites on mineral surface. Thus, in soil with Pi deficiency and with great quantity of clay mineral and oxides, the adsorption of Pi is higher, resulting in high rates of P-fertilizer to supply plant requirement [19, 23], which is P-fertilizer the majority source of Pi in highly weathered soil.

Iron and aluminum oxides are considered the majority constituents of clay fractions in the most Brazilian soils cultivated with forages, example the order of Latosols (oxisol) that cover approximately 65% of national territorial, where 50% of this area belongs to Cerrado biome [24, 25], which are explored by livestock and grain crops and are considered the highest world celery [26]. The oxides are the most effective in P-adsorption [27]. Among the oxides, goethite are considered the most effective to bind orthophosphate, consequence of facility to bind in OH^- groups in mineral surface in the complex of adsorption in external sphere, beside the crystal morphology and higher surface area in relation to hematite [23].

Gibbsite is the other oxide related to Pi adsorption, however with lower effectivity in relation to goethite. In highly weathered soil, gibbsite has higher amount in clay fraction, and its total capacity of P-adsorption can be above the Fe-oxides [22]. On the other hand, in kaolitic soil, the lower amount of gibbsite can decrease the capacity of orthophosphate adsorption, due to its lower bind sites on mineral surface. In oxisol, the preview liming can decrease Al and Fe content due to precipitation, consequently decreased the adsorption sites. Conversely, excess of liming can increase the Pi associated with Ca (P-Ca), which decreases the Pi availability.

Therefore, the relative proportion of Pi compounds with Fe, Al, and Ca is dependent of soil pH, as well as type and quantity of mineral existent in clay fractions [23]. Thus, the pH also affects the chemical forms of Pi found in soil, following the dissociation of H_3PO_4 [28]:

$$H_3PO_4 \rightarrow H^+ + H_2PO_4^- \quad \text{Log K1} = 2.12 \tag{1}$$

$$H_2PO_4 \rightarrow H^+ + HPO_4^{-2} \quad \text{Log K2} = 7.20 \tag{2}$$

$$HPO_4^{-2} \rightarrow H^+ + PO_4^{-3} \quad \text{Log K3} = 12.33 \tag{3}$$

The majority of soil cultivated in Brazil shows pH range of 3–6 units, the predominant Pi form is the $H_2PO_4^-$, this way, $H_2PO_4^-$ are considered more available to plants due to soil acidity [29]. The HPO_4^{-2} is less available in soil due to its higher capacity to bind with Fe and Al than $H_2PO_4^-$, thus increasing in soil pH above 7 decrease the availability of P in soil, besides the P-Ca precipitation. In pasture managed, P-fertilizer periodic and adequate stocking rates can increase the available Pi in soil (form 7 to 18 mg dm^{-3}) in 0–10 cm depth, consequently increasing in forage yield (from 3 to 7.4 Mg ha^{-1}) [20].

The soil class is the factor that has more impact on Pi forms, while the conditions of use and soil tillage control the content of total Po, as the Po in microbial biomass and activity of phosphatases, which means that there are mineralization/immobilization of P from organic compounds and solubilization/precipitation of inorganic phosphates [19, 20]. The actual knowledge about organic and inorganic P in soil restricts our capacity of developing management strategy to promote the use of more efficient P-fertilizer in crop production. However, in forage grass periodically managed, there is improvement in physical quality, by establishment of soil structure and chemical through different sources of Po, resulting in affect the biological activity and enzymatic response by bio-availability of P, occasioning in stimulation of specific mechanism of mobilization of labile P, returning the recalcitrant P back to the cycle [21].

4.1. Microbial activity in phosphorus dynamics

Phosphorus dynamics in soil are associated with environmental factors that control the microbial activity, which immobilize and release orthophosphate ions and the physical–chemical properties and soil mineralogy [30]. Mineral (Pi) and organic P (Po) forms are the majority pools in soil, since in oxisol, there is predominance of Pi bonded with high energy in inorganic fractions, and Po forms show stabilized forms, physically and chemically. According to the energy of binding, the Pi can be classified as labile and nonlabile. In this context, the labile fraction is shown by group of phosphate compounds capable to replace the soil solution, when orthophosphate Pi are uptake by plants and soil microorganisms [31].

Thus, the microorganisms show important rule in immobilization and availability of Po, especially in the function of enzymatic excretion that acts in this process. Phosphatases represent a wide group of enzyme that catalase the hydrolase ester and anhydrous of phosphoric acid, as well as showing positive and significant correlation with microbial biomass carbon (C-MBC) and Po of the soil microbial biomass (P-SMB) [32]. Phosphatases can be excreted by microorganisms or root of higher plants [33]. The intensity of excretion by roots and microorganism is determined by your orthophosphate demand that is affected by soil pH. Soil enzymatic activity and biochemical reactions can be a sensitive indicator of soil quality and stressed conditions in soil promoted by soil tillage [33].

The soil tillage and use can modify the activity of enzymes in soil and consequently the availability of Pi in soil. The pH increasing in soil due to liming broadcasting can stimulate the activity and diversity of microorganism population, resulting in increasing the enzymatic activity and consequently affect the nutrient cycle. Except the acid phosphatase, many enzymes have your enzymatic activity favored by pH increasing [33]. The transformations of P in soil are majority guide by microorganisms associated by a combination of factors including the plant species, type of soil, and environment [34]. The exudation of different plant species results in stimulation of different microbial species in rhizosphere via root exudation, which include sugars, aminoacids, organic acids, hormones, and vitamins [34].

In terrestrial ecosystems, the root exudation represents 40% of organic compounds [35]. The amount of enzyme exudates depends on difference in carbon metabolism among plant species under different growth stage [32]. Plant species and growth stage determine the effect of microbial activity and C-SMB, as well as the mineralization of the nutrients in dry matter composition. It is important to know that Po associate with soil microorganism is a source of P labile for plant uptakes after mineralization.

Under conservationist system, the effects of plants on P availability (P-labile) differ among the species of crop rotation, thus with the increasing of root exudation and microbial activity in rhizosphere, there are higher levels of soluble carbon, which contribute to increase the microbial activity and biomass, resulting in increasing of utilization and solubilization of Po and Pi [32].

The mineralization of Po through soil microbial depends on carbon availability in soil. It has already observed lower amount of C-SMB in soil with low fertility and that the Po mineralization is limited in soil with low amount of carbon [36]. Nevertheless, in soil with low Pi content was not observed inhibition of C-SMB growth, but Pi content in soil affects the amount of immobilization by microbial biomass [37].

Under restrictions of Pi content in soil, the activity of phosphatase increases, resulting in higher mineralization of Po [38]. Besides the plant species and amount of available carbon, the quantity of immobilization and mineralization is dependent of microbial composition in soil [36]. The microorganism uses organic compounds phosphorylates as carbon source, resulting Po mineralization. The procedure can increase Pi content by plants in locations where P is limited [36]. In conditions where Pi is supplied through P-fertilizer in enough amounts for plant and microorganism requirement, the Po mineralization decrease due to phosphatase inhibition [38]. On the other hand, in conditions where Pi is not enough in soil, the activity of phosphatase increases. This fact is associated with the capacity of plant and soil microorganism having responded to alteration in environment for acquisition limited nutrient in soil, as the case of plants and microorganisms that increase phosphatase synthetize in constraint content of Pi in soil [39]. In the study accomplished by [38], decrease in microbial biomass in function of Pi increment in soil was not observed; however, the reduction of 10% in mycorrhizae fungi was observed. These results were sustained by decreasing in carbon by microorganism under condition of enough Pi in soil, so the occurrence of mycorrhizae fungi decreases.

4.2. Phosphorus forms in soil

The total content of P in soil is distributed under different degrees of lability. Its lability, which are established through the linking energy with colloids, defined the labile forms, moderated labile, and no labile [40], define this distribution of P in soil. In each P form, the content of P is variable in function of soil type, tillage, cultivation, climate, and content of organic matter. The classification of P forms in soil according to lability is such arbitrary because depending on the P uptake rate by plants, the P forms are defined as labile or no-labile [41]. The division of P forms in soil is necessary to understand that there are pools of P in soil with higher or lower capacity to supply soil solution [42].

In soils under natural environments, the content of P in soil depends on primary material, which affects many physical–chemical soil properties due the difference in textural features, chemical composition, and mineralogy fractions of soil. In highly weathering soil and natural environment, the Po compile in important fraction of total P, most of labile P, which are converted in Pi through organic matter mineralization. However, under cropping system with P-fertilizer, there are accumulation of Pi and Po with different degrees of bonding energy, though the accumulation is more pronounced in Pi forms [43].

Changes in P forms distribution in soil can also be associated to system of soil tillage, P exported by harvesting, rates of P-fertilizer reposition, P-source applied, and ability of plants in using P reserves in less labile fractions [44]. In general, under tropical soils, there is predominance of Pi forms than Po [45–47]. Organic P shows great importance because its majority part is labile and moderate labile P fraction, which acts in supply soil solution with Pi when its concentration decreases through plant uptakes and microorganism immobilization [45, 48]. Phosphorus immobilized in microbial biomass constitutes a potential reserve of P able to supply forages in absence of P-fertilizer.

5. Phosphorus labile

In the Pi labile fraction, the P is bonding with less strength to soil colloids through monocordenates linked, what allows desorption of nutrients to soil solution when Pi content decreases soil solution. The Po labile is the proportion of Po associated with the organic materials of easy mineralization. Thus, P labile is the amount in balance with soil solution and represents the group of compounds capable to supply soil solution content that is available for plant uptakes, which are dependent of weathering degree, soil mineralogy, granulometry, SOM, physical-chemical properties, biologic activity, and predominant vegetation [49]. In soil with low P-labile content, the majority strategy to increase P content is through P-fertilizer, by organic or inorganic forms.

6. Phosphorus moderated labile

The moderated labile P forms are represented by P linked to chemical sorption to Fe and Al oxides and clay. These mineral are presented in soil and because its capacity in form complex with high energy, the sorption is slow and can occur in medium to long term. Inorganic P forms are physically protected in inner surface of soil aggregate and linked to Ca (parent material) and are considered P moderated labile [49], besides Po in the SOM stable.

7. Phosphorus no labile

Phosphorus no labile (recalcitrant) is the P linked with higher energy in soil and is strongly adsorbed or precipitate in insoluble compounds [20, 50]. The no-labile Po fractions are associated to humic compounds and physically protected inner of microaggregates. The no-labile P forms are represented to organic and inorganic recalcitrant compounds, which orthophosphate linked energy is through two coordinate links, this double link does not allow immediate desorption of P. In order to optimize P-fertilizer as growth factor for plants, the no-labile P must be quantified, understanding and controlled majority for highly weathering soil. The properties and mineral constituents of clay fractions are responsible by speed of transformation of labile-P through no labile-P [51].

8. Determination of P forms

In order to better understanding of P dynamic in soil, the knowledge of different P fractions in soil is prerequisite, which can be accomplished through sequential extraction by different extractors [52]. Thus, with the determination of P fractions, it is possible to have valuable information about P availability in soil. The relation of P forms in soil and its distribution is a relevant question, and the use of chemical fractionation technic that determine the quantity and distribution of P forms in soil is valuable in the understanding of P dynamic under different agroecosystems. There are many schemes of soil P fractionation, which were combined by [53] in the following classes: fractionation to Pi forms, fractionation to Po forms, fractionation to Pi and Po, and fractionation to Pi, Po, and P microbial.

Many authors have been using the fraction proposed by Hedley [44, 54–56]. Hedley's fractionation is widely used [57], majority in researches about P dynamic and cycling in soil associated with primary material, soil tillage, and use into diverse crop systems. Hedley's fractionation allows the determination of Pi and Po in soil based on chemical extractors with increasing capacity of P extraction; however, Hedley's fractionation only cannot explain the P forms in soil. Thus, the work of Cross and Schlesinger [49] correlated the P forms in soil with the Hedley's fractionation, resulting in separation of the labile P from the no-labile P forms, which shows the possibility of identifying the preferential forms that P are retained in soil.

Assume that the quantification of P forms in soil can be accomplished through labile P, which are composed by the sum of P extracted by anionic exchange resin and fractions of Po and Pi extracted with $NaHCO_3$. The moderated labile P fractions are composed by Pi and Po fractions linked with higher energy to Fe and Al, which correspond to P extracted with NaOH (0.1 and 0.5 mol L^{-1}). The no-labile P fractions are composed by fraction extracted with HCl (P fraction linked to Ca with constraint availability) plus residual P (Po and Pi fractions insoluble) [51].

Phosphorus forms and degrees of lability in soil change with the soil properties. In new soils and with lower degrees of weathering, Ca phosphates are the major supplier of P to life organisms. Conversely, in highly weathering soil, the bio-cycling of organic phosphates has great importance in maintenance of bioavailability, throughout it is not enough for maximal economic yield of commercial crops [31]. Phosphorus dynamic in natural ecosystems and managed are established majority by interactions of nutrients with the Pi and Po pools and with soil microorganism. Thus, researches about P dynamic and availability require separation and identification of different P fractions in soil. The technic of P fractionation aims to identify the preferential fractions, which P are linked in soil, and your occurrence and magnitude that the fractions have to supply plant P requirement. Thus, studying the P fractions is essential because of the great difference between many types of soil and crop systems, much more about the practices of liming and fertilizer application technics, which alter the P dynamic in soil [42].

8.1. Phosphorus use efficiency

The low availability of phosphorus (P) in Brazilian soils requires the application of this nutrient in grain, fiber, wood, horticulture systems, and forages, mainly via soluble inorganic fertilizers. However, the imminent depletion of phosphate rock reserves in the next century [58], and the use of P rates greater than the ability of the soil to retain can make it polluting in water, make the use of phosphate fertilizers to be minimized and used more efficiently. The low availability of P is one of the most restrictive factors for livestock, since the forage plants can be very demanding in P, due to a higher production of biomass, consequent to the greater extraction and export of this nutrient [12].

The nutritional adjustment in crops depends, beyond the technology level used, on the ability to uptake and P use by plants, characteristic that is related to morphophysiologic parameters of genetic orientation specific of each cultivar, and expressed in function of environmental conditions in the cultivation area. Evaluating biomass production and agronomic efficiency as function of phosphorus supply in different genotypes of *Brachiaria brizantha*, [59] observed that the Arapoty variety and the B5 genotype showed greater efficiency in the use of P, because it produced more shoot biomass per unit of applied P, being able to indicate for breeding programs of this species, while the Capiporã variety and the B12 genotype were less efficient.

According to [60], efficient plants in phosphorus use have genes that confer adaptive mechanisms to contour low availability of nutrients in the environment, among them modifications in architecture and growth of the root, increase in phosphatase production, and change in the activity of several enzymes in glycolytic route. Nevertheless, in presence of appropriate nutritional P levels, these genes may not express themselves, resulting in a lower plant response to the environmental improvement. Thus, differences between genotype plants in relation to efficiency in the use of this nutrient can be assigned to the fact that the absorption of the phosphorus present in the soil solution occurs via root interception, so that plants of bigger root system present advantages in its capture [61].

Some plants have different abilities to remove and absorb nutrients from the soil, mainly by processes occurring in the rhizosphere. For P, they use different mechanisms to access the less labile soil P forms and favor the cycling of P in the system as: increase in root/shoot ratio, root surface or increase in absorption rate per root unit [62]; increase in the number, shape, and thickness of the root hairs [63]; root phosphatases exudation [64] or organic compounds capable of complex metals phosphate associated [65]; by mycorrhizal association in which the fungal hyphae extend the root area [66] or with other microorganisms capable of favoring the cleavage or breaking of organic compounds with the consequent release of the phosphate anion [62, 67].

The detection and possible exploration and use of genotypic plant differences for P efficiency is one of the viable strategies to reduce the problem of P deficiency in tropical and subtropical regions, as a consequence of the naturally low P levels and the high capacity of fixation in soils. The development of plant genotypes adapted to the adverse conditions of soil fertility, notably to phosphorus deficiency, the introduction of selected material to certain environments are interesting aspects from the point of view of the efficiency in the P-fertilizer use and the sustainability of the productive system.

Together, the cycling of P by the plants is also important, because these have different degrees of adaptation to access the soil P. There are plant genotypes that take advantage of inorganic phosphorus (Pi) by their roots or associations with mycorrhizal and those that use organic phosphorus (Po) by specialized enzymatic mechanisms, for each type of phosphate esters, which are used as nutrient sources [68]. As these mechanisms vary with plant species, in order to optimize soil P use by plants in agricultural systems, it is essential to identify those with the greatest potential to absorb and cyclize soil P, especially those that can be used commercially.

The interaction between fungus and plant varies according to the genotype, since they have affinity for root systems with characteristics that favor mycorrhizal symbiosis, like higher exudation of lipids, carbohydrates, and carbon compounds. These studies shown that low doses of P increase mycorrhization and efficiency of mycorrhizal fungi in promoting dry matter increase; however, high doses of this nutrient affect negatively the mycorrhization [69]. Evidence suggests that this is caused by a reduced reliance of plants on arbuscular mycorrhizal fungi for phosphorus, which is concomitantly dependent on N availability such that a reduced N:P ratio can suppress arbuscular mycorrhizal fungi [70]. The efficient cultivars in phosphorus use can be related to higher levels of mycorrhizal interaction and consequent increase in nutrient absorption.

According to [71], for managing grazing systems for improved P-use, efficiency should be to avoid over-application of P ('P equilibrium fertilization' practices in which P input in manure and fertilizer does not exceed P output in products); use pastures that are productive at lower plant available P concentrations; legume-based pasture systems; and plant traits that address P-balance efficiency (more root foraging, favorable root architecture, high specific root length, long root hairs, root adaptation to P stress, and high root growth rates) [71].

For pastures that are productive at lower plant-available P concentrations; legume-based pasture system, the key legume species in these systems (e.g., subterranean clover (*Trifolium subterraneum* L.) and white clover (*Trifolium repens* L.) often have coarse roots and short root hairs and have higher critical P requirements than the forage grasses with which they are grown. The pasture is fertilized to meet the higher P requirements of the legume, because legume N-fixation drives overall productivity. It will be necessary to find legumes with lower critical P requirements to improve the P-balance efficiency of these pasture systems. Temperate pastures differ from some mixed pastures grown on infertile acid soils of the tropics (e.g., *Stylosanthes capitata* Vogel, *Zornia latifolia* Sm.–*Brachiaria decumbens* Stapf., and *Andropogon gayanus* Kunth grasslands of Central America).

9. The 4R's nutrient stewardship' for P-fertilizer in forage grasses

The concept of best management practices (BMPs) for fertilizer application is universal used for grain crops, following the 4R's nutrient stewardship (right source, right rate, right place, and right time) [14] (**Figure 1**). All the BMPs that follow the 4R's nutrient stewardship must be associated with environmental, social, and economic benefits (**Figure 1**). Despite the chapter being focused on P-fertilizer, all fertilizers must follow these concepts in order to improve sustainable agriculture. First, this concept of 4R's nutrient stewardship was introduced by [14], which was followed and applied worldwide with great acceptance in many

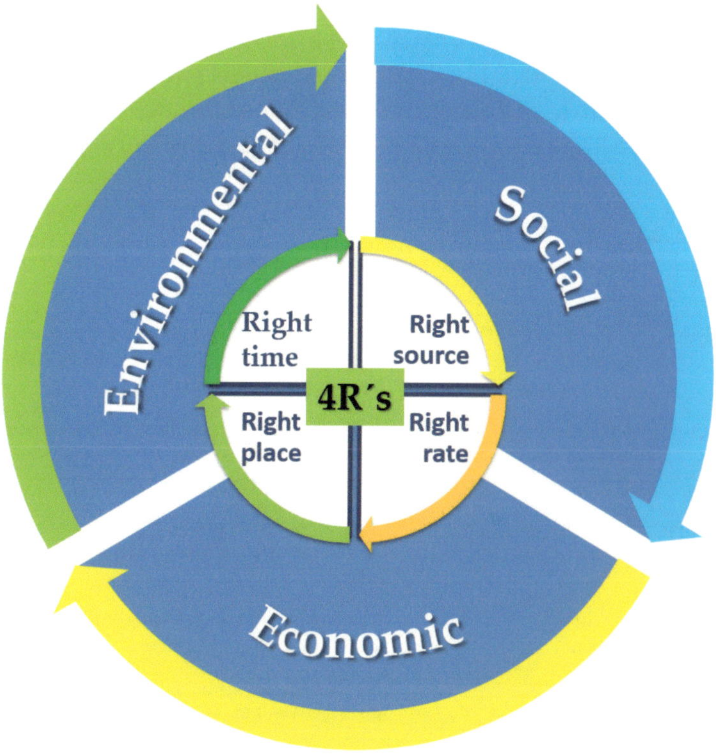

Figure 1. The 4R's nutrient stewardship for P-fertilizer on pastures. Adapted from Kochian [5] and elaborate by the authors.

crops. The application of 4R's in forages can be in advantage in BMPs in tropical region, due to many wrong practices applied in grassland related to nutrients management. The natural P-fertilizer sources are finite, which justify its use carefully to improve it efficiency. Forage grasses (*Brachiaria* spp.) are recognized to be very efficient in P use efficiency [5], nevertheless in tropical region, P availability in soil is very restricted.

9.1. Right source

The most P-sources used in forage grasses are soluble P-sources (**Table 1**), and less is relate to reactive natural phosphate (RNP), even with such researches showing the incorporation of RNP can result in approximated efficiency than soluble sources as triple phosphate [72]. Even with low solubility, maybe in sand soil, it is possible to improve the uptake of forage through the use of RNP in pasture because of minor adsorption site than in clay soil. Due to RNP be slow release P-source, the capacity of P-sorption may overcome the release, resulting in P deficiency for plants [50].

Even with the recommendation for RNP application be in soil with low pH (more acid), in some cases with pH_{CaCl} = 6.0 was possible to observe increasing of forage volume in 25% for *Panicum maximum* cv. Massai [73]. In the case of *Brachiaria decumbens*, application of RNF promoted 30% of dry matter production with RNF was applied [72]. Therefore, it is important to remember that acidity in rhizosphere is higher than surround soil [15], which can contribute to improve RNF solubility and plant uptake. The evidence of possible use of RNF in pasture as P-source is quite clean, but their use in pastures is not yet widely applied. Reactive natural phosphate can be an alternative to decrease the use of P-source more soluble used in pastures nowadays, which are in jeopardy due to limited P-source in the world (**Table 1**).

P-fertilizer source	Minimum guarantee*	
	Nutrient content and form	Nutrient solubility
Simple superphosphate	18% of P_2O_5, 19% of Ca, 11% of S	Total content of P_2O_5 content soluble in ammonium neutral citrate plus water and minimum of 16% soluble in water. Total content of Ca and S.
Triple superphosphate	48% of P_2O_5 10% of Ca	Total content of P_2O_5 content soluble in ammonium neutral citrate plus water and minimum of 36% soluble in water. Total content of Ca.
Diammonium phosphate (DAP)	17% of N 45% of P_2O_5	Total content of N and P_2O_5 content soluble in ammonium neutral citrate plus water and minimum of 44% soluble in water.
Monoammonium phosphate	9% of N 48% of P_2O_5	Total content of P_2O_5 content soluble in ammonium neutral citrate plus water and minimum of 44% soluble in water.
Reactive natural phosphate	27% of P_2O_5 28% of Ca	Phosphorus determined as P_2O_5 total and minimum of 30% of the total soluble in citric acid at 2% in relation 1:100.

*Source: Adapted from Ministry of Agriculture, Livestock and Food Supply (MAPA) [75].

Table 1. Major P-sources used in forages, specification of the simple solid sources of P with minimum nutrient content guarantee by law.

Application of P-fertilizer sources with higher solubility can improve biomass production of pasture, however in soil highly weathering the precipitation of P-Fe and P-Al can result in no-labile P forms and consequent decrease of P-fertilizer use efficiency. Thus, application of P-source can be wisely decided to improve phosphorus use efficiency (PUE) and decrease P-sorption in soil.

The use of monoammonium phosphate (MAP) as P-sources in pasture can result in lower PUE in comparison to simple superphosphate. Monoammonium phosphate is more soluble and its use in pasture as P-source can result in faster P-sorption in highly weathering soils with higher buffering capacity. Then, most of P-fertilizer applied is going to be fixed in soil through time and PUE tends to decrease. The price of P-source normally determines its use for farmers, but the P-source must be decided directing to more efficient P-source that long lasting are going to obtain financial returns associated with absence of environmental negative impact. However, higher soluble P-sources tend to increase the dry matter production in short term, as the case of triple superphosphate in comparison to RNP in *B. decumbens* and *B. brizantha* [74].

9.2. Right rate

In order to define the P-fertilizer rate in pasture, it is necessary to know the plant requirement, because the genotypes show different demands for P acquisition. The genotypes *Panicum maximum* is very demanded for P, followed by *Brachiaria brizantha*, *Brachiaria ruziziensis*, *Brachiaria decumbens*, *Brachiaria humidicola*, and *Stylosanthes* spp. (**Table 2**). The degree of P-requirement for pastures species and soil P content determines the P-fertilizer rates to be applied. The difference in forage species in P-requirement is so evident that it is crucial to separate the species by the groups of P requirement (**Table 2**). In **Table 2**, the degree of P requirement for forages species which is very used in Cerrado region to direct P-fertilizer rates recommendations, which was developed by [76] are shown.

Forages species	Degree of P requirement
Stylosanthes spp.	Low
Andropogon gayanus cv. Planaltina	Low
Brachiaria decumbens	Low
Brachiaria ruziziensis	Low
Brachiaria humidicola	Low
Paspalum atratum cv. Pojuca	Low
Brachiaria brizantha cv. Marandu, Xaraés, Piatã, Ypiporã	Medium
Panicum maximum cv. Massai	Medium
Panicum maximum cv. Mombaça, Tanzânia, Aruana, Tobiatã	High

Source: Adapted from Martha et al. [76].

Table 2. Degrees of P requirement for some forages species.

The quantification of Pi availability in soil is the first step to define the right rate of P-fertilizer. As reported previously, tropical region is composed by poor soil with constraints in P availability for plant growth. In the most Brazilian soils, the absence of P-fertilizer avoids plant normal development and economic yield.

The amount of P-fertilizer is very dependable of soil clay content, thus in sand, soil P-fertilizer rate must be lower than clayed soil due to less buffering capacity. In order to recommend P-fertilizer rates, the P extractor plays an important role in this definition. If the extractor is Mehlich-1, the clay content in soil must be determined to interpret the P content availability in association with degree of P requirement by forage species (**Tables 2** and **3**). With the increase in clay content, the amount of available P decreases, result of higher buffering capacity of P, than the amount of P-fertilizer is higher in clay soil than in sandy soil. When P-resin is used, the extractor has no significant dependence to clay content [51], thus quantification of clay is not necessary to define the P level.

The amount of P-fertilizer to achieve inadequate P content can be easily determined by the following formula: P-fertilizer rate (kg ha^{-1}) = [(P expected content−P available in soil) × P buffering capacity] [76].

9.3. Right place

First of all, it is quite important to define the moment of P-fertilizer in forages; (1) forage implementation and consequently P-fertilizer correction for establishment, and (2) forage maintenance that is recommended to remain the adequate P level in soil. The implementation of forage can be conducted under no-till and tillage system, which change completely the P-fertilizer placement. Usually, in tillage soil the P-fertilizer in broadcasting and incorporate with arrow disc into 20 cm depth. This procedure is essential to improve the forage root

Interpretation of P content in soil (mg dm^{-3})—Method of Mehlich-1				**P buffering capacity**	
Clay content (%)	**Degree of P requirement**			**Mehlich-1**	**Resin**
	Low	**Medium**	**High**	**(kg P_2O_5 ha^{-1})/(mg dm^3 of P)**	
≤15	>9.0	>11.0	>14.0	5	6
16–35	>7.0	>9.0	>12.0	9	9
36–60	>4.0	>5.0	>6.0	30	14
>60	>2.0	>2.5	>3.0	70	19
Interpretation of P content in soil (mg dm^{-3})—Method of anionic exchange resin extractor (P-resin)					
Degree of P requirement					
Low		**Medium**		**High**	
>7.0		>9.0		>12.0	

Source: Adapted from Martha et al. [76].

Table 3. Phosphorus content in soil interpretation through critical levels of adequate P content in 0–20 cm depth by Mehlich-1 and resin methods, based on soil P adequate content and plant requirement for pasture establishment.

acquisition for P, which should be consider due to slow mobility of P in soil that is almost all by diffusion [2]. However, in soil with high P-sorption to incorporate P-fertilizer in soil can decrease its availability.

In integrated crop-livestock system, the no-till system P-fertilizer tends to be applied deeper in soil (10–20 cm) in the grain crop production instead of application directly for forage; however, in integrated crop-livestock system, the content of P in soil usually are above 5 mg dm^{-3}, which is considered enough for *Brachiaria* spp. requirements. It happened because soybean or corn has higher P requirement than forages, thus the residual P-fertilizer in soil is enough for forage growth.

The maintainace of P content through time under stablished pasture is usually proceeded broadcasting P-sources on soil surface without deep incorporation, this procedure has shown quite efficient to remain forage in adequate growth for decades.

9.4. Right time

The right time of nutrient application is related to the nutrient uptake pattern; in case of P, forage demand higher amount of P in establishment and vegetative growth. Then, correction of P content in soil before sowing is decisive to obtain faster initial growth, and remaining P content in soil will maintain the biomass production. Pasture implementation and maintenance are the two moments for P-fertilizer application. During maintenance, it is quite fundamental to remain the P content above the critical limit for each forage species nutritional demand (**Table 3**).

Phosphorus fertilizer to correct the P levels in soil is usually done in forage implementation and its content are maintained through vegetative growth. The recommendation to improve PUE is the application of P-source after soil acidity correction, especially in highly weathering soil.

Rain after broadcasting P-fertilizer in forages can cause surface runoff and soil erosion, which must be taken into consideration to decide the right time to apply P-fertilizer. Phosphorus has low mobility in soil, results in high concentration in soil surface [15], when P-fertilizer are broadcasting on soil surface in forage grasses.

10. Soil nutrient interactions with phosphorus

Principal component analyses (PCA) were performed to illustrate the relationship between P available in soil with some chemical and physical soil properties (**Figure 2**). Most of the variables were attributed to two principal components (PCs). The most PCs loadings were significant based on selection criterion defined by [77]. The two first PCs combined explained 53.35% of the whole variability in database. The first PC was positively correlated to base saturation (BS), exchangeable Ca + Mg, Ca saturation, exchangeable Ca, exchangeable Mg, Mg saturation, pH_{-CaCl}, pH_{-H2O}, exchangeable K, S, $P_{-Mehlich}$, Ca/Mg, clay, organic matter, P_{-resin}, and negatively correlated to H + Al saturation, exchangeable H + Al, sand, Fe, Al, Al saturation (**Figure 2**). The second PC was positively correlated to exchangeable K, sand, exchangeable Mn, K saturation, exchangeable Fe, and negatively correlate to clay, Ca/K (**Figure 2**).

The structures obtained by PC-1 and PC-2 are supported by some rules relative to interaction between nutrients in soil, but other structures should be studied carefully for a better understanding (**Figure 2**). The opposite direction between $P_{-Mehlich}$ and P_{-resin} with exchangeable Al is clean evidence of the antagonism between these elements in soil, as exposed before, the soluble Al in soil can precipitate P (P-Al), which results in less P available in soil for plant uptake, consequently, lower exchangeable Al in soil to increase P availability.

Phosphorus-Ca, P-Fe, and P-Al are forms of P precipitation, and its solubility is associated with soil pH. The soil acidity correction is a practice to improve PUE, resulting in more P available due to Al precipitation before P-fertilizer application in 0–20 cm depth when liming is incorporate. Liming application is quite useful in soil with high amount of Al and with pH below 4.7 units. The increasing in soil pH above 4.7 can decrease soluble Al in zero (**Figure 3**). Nevertheless, for a better PUE, it is important to have no limitation in other nutrients availability in soil, as the case of N, S, K, and other essential nutrients. The soil compaction in clayed soil with animal trampling is a problem that can decrease the PUE due to impossibility of root P acquisition. As observed in **Figure 3**, exchangeable Ca + Mg were in opposite direction in first PC with Al, Al saturation, and exchangeable H + Al, which is possible to infer that the increasing in exchangeable Ca + Mg in soil are associated with liming, and consequently, its application tends to decrease soil acidity above pH = 4.7 units and Al saturation are totally precited in

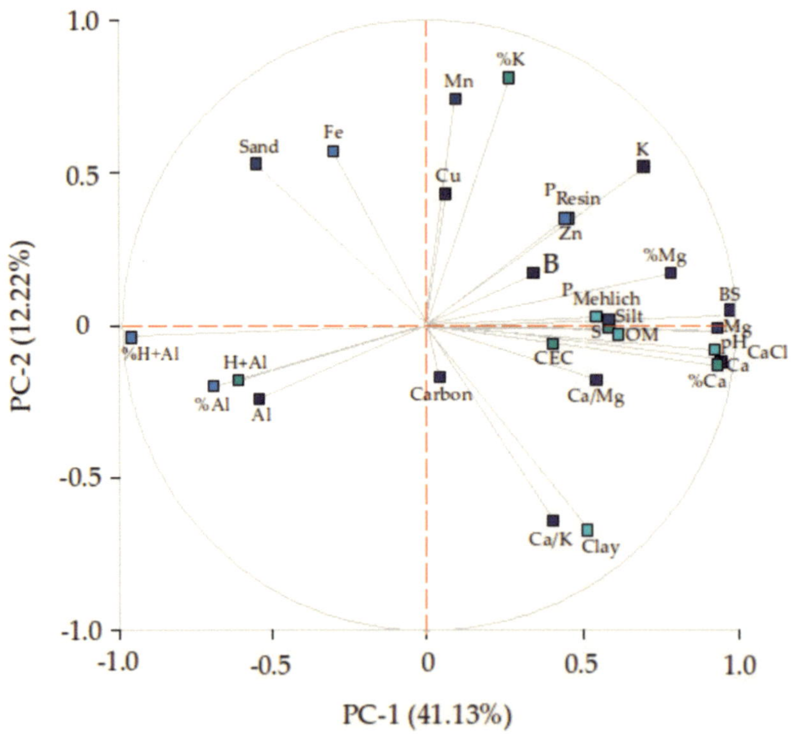

Figure 2. Monoplot of principal component analysis (PCA) for some physical and chemical soil properties under pastures in Cerrado West region of Bahia State, Brazil. PC—principal component; PC-1—the first principal component; PC-2—the second principal component; exchangeable Ca, Mg, K, Fe, Mn, Al and Zn.

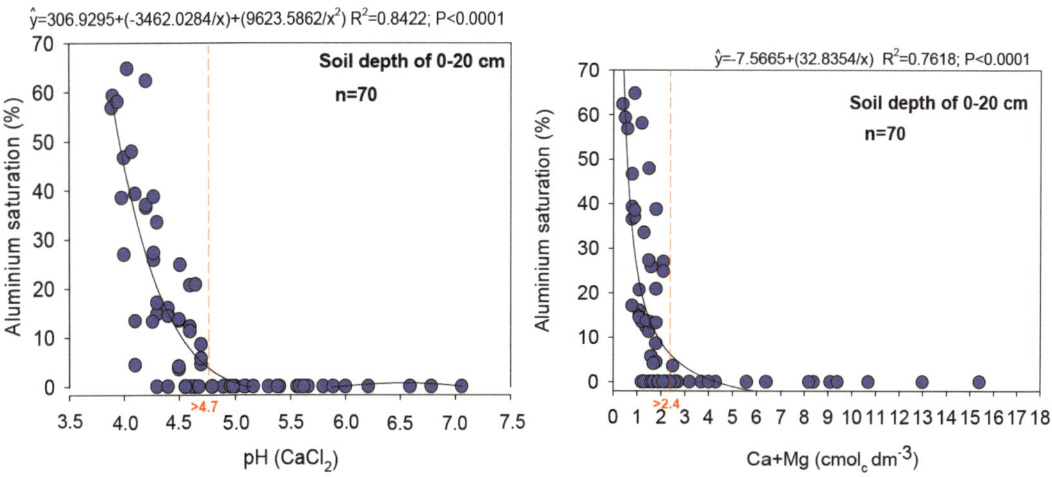

Figure 3. Relation between soil pH and Al saturation from soil samples collected in oxisol from 0 to 20 cm depth. Source: Authors, data not published previously.

hydroxides (**Figure 2**). Conversely, the decreasing in Al soluble results in less site to fix orthophosphate in soil and tends to increase its availability; however, it is quite important to mention that overrate of liming can result in Ca-P precipitation and decrease P availability as well.

For Phosphorus labile and P no-labile, it is quite important to observe that over time, 1 year after P-fertilizer, 79–95% is turned into no-labile P [51]. Thus, the P-fertilizer must be well observed in order to avoid decreasing in PUE. Soil organic matter is a way to improve P availability from P no-labile fractions. Thus, increasing SOM can be in alternative to improve PUE in tropical soils. The importance of using P-resin extractor instead of Mehlich-1 in soil with historic of P-fertilizer applied with reactive natural phosphate (RNP) is because the effect in acid extractor as Mehlich-1 in solubilize the P linked to Ca in RNP, resulting in overestimation of P content in soil [51].

It is not easy to improve PUE in soil with high P-sorption capacity. Most of Brazilian soils are located in region with very high P buffering capacity, resulting in low P availability [5]. The ability of forages as *Brachiaria* spp. to yield well with lower extractable soil P content is primarily associated with morphological traits such as long fine roots and long root hairs that enable foraging for available P and its uptake from soil solution. Together, these traits confer a large root hair cylinder volume (RHCV) which is strongly correlated with P uptake [78]. Physiological root traits, such as exudation of carboxylates (low-molecular-weight organic anions) and phosphatases into the rhizosphere, can potentially enable plants to these sources of soil P [71].

11. Concluding remarks

Tropical soils are quite limited in P content, due to the natural formation with parent material poor in P content and weathering. Thus, in order to improve phosphorus use efficiency, 4R's must be followed and adjust in site-specific conditions. Improving PUE in highly weathering soil with high buffering capacity is a challenge; however, all the combination of best management practices for P-fertilizer application can result in better use efficiency. Based on the scarcity of natural P sources in the whole world, the use of alternative P-sources should be incentivized, and more researches about this issue are needed for better understanding in forages.

The propose to direct the knowledge in P dynamic with the best management practices can be a useful tool to improve P-fertilizer efficiency in forages in tropical soil highly weathered. The recognition and potential examination and use of genotypic forages plants with higher P use efficiency is a sustainable approach to aim the problem in tropical soil with higher P-sorption capacity. The development of plant genotypes adapted to the adverse conditions of soil fertility, notably to phosphorus deficiency, the introduction of selected material to certain environments are interesting aspects from the point of view of the efficiency in the P-fertilizer use and the sustainability of the productive system.

Author details

Ademar Pereira Serra[1*], Marlene Estevão Marchetti[2], Elisângela Dupas[2], Carla Eloize Carducci[2], Eulene Francisco da Silva[3], Elaine Reis Pinheiro Lourente[2], Alessandra Mayumi Tokura Alovisi[1] and Simone Candido Ensinas[4]

*Address all correspondence to: ademar.serra@embrapa.br

1 Brazilian Agricultural Research Corporation (EMBRAPA), Campo Grande, Brazil

2 Federal University of Grande Dourados (UFGD), Dourados, Brazil

3 Federal Rural University of the Semi-arid Region (UFERSA), Mossoró, Brazil

4 State University of Mato Grosso do Sul (UEMS), Cassilândia, Brazil

References

[1] Whitehead DC. Nutrient Elements in Grassland: Soil-Plant-Animal Relationship. Oxford: Cabi Publishing; 2000 363 p

[2] Marschner P. Marschner's Mineral Nutrition of Higher Plants. 3rd ed. London: Elsevier; 2012 643 p

[3] Bushinsky DA, Myles W. Mineral metabolism: Progress in the face of complexity. Current Opinion in Nephrology & Hypertension. 2016;**25**:269-270. DOI: 10.1097/MNH.0000000000000242

[4] Food and Agriculture Organization of The United Nations (FAO). Rome, 2011. Current World Fertilizer Trends and Outlook to 2015. Available from: http://www.fao.org/3/a-av252e.pdf

[5] Kochian LV. Plant nutrition: Rooting for more phosphorus. Nature. 2012;**488**:466-467. DOI: 10.1038/488466a

[6] Louw-Gaume AE, Rao IM, Gaume AJ, Frossard E. A comparative study on plant growth and root plasticity responses of two *Brachiaria* forage grasses grown in nutrient solution at low and high phosphorus supply. Plant and Soil. 2010;**328**:155-164. DOI: 10.1007/s11104-009-0093-z

[7] Euclides VPB, Valle CB, Macedo MCM, Almeida RG, Montagner DB, Barbosa RA. Brazilian scientific progress in pasture research during the first decade of XXI century. Revista Brasileira de Zootecnia. 2010;**39**:151-168

[8] DIEESE – Departamento Intersindical de Estatística e Estudos Socioeconômicos. Estatísticas do meio rural 2010-2011. In: Paulo S, editor. 4. NEAD, MDA: DIEESE; 2011

[9] Freitas ME, Souza LCF, Salton JC, Serra AP, Mauad M, Cortez JW, Marchetti ME. Crop rotation affects soybean performance in no-tillage system under optimal and dry cropping seasons. Australian Journal of Crop Science. 2016;**10**:353-361

[10] Lourente ERP, Silva EF, Mercante FM, Serra AP, Peixoto PPP, Sereia RC, Ensinas SC, Marchetti ME, Neto Neto AL, Alovisio AMT, Cortez JW. Agricultural management systems affect on physical, chemical and microbial soil properties. Australian Journal of Crop Science. 2016;**10**:683-692

[11] Ensinas SC, Serra AP, Marchetti ME, Silva EF, Lourente ERP, Prado EAF, Matos F, Altomar PH, Martinez MA, Potrich DC, Conrad VA, Jesus MV, Kadri TCE. Cover crops affect the soil chemical properties under no-till system. Australian Journal of Crop Science. 2016;**10**:1104-1111

[12] Oliveira PPA, Oliveira WS, Corsi M. Efeito residual de fertilizantes fosfatados solúveis na recuperação de pastagem de Brachiaria brizantha cv. Marandú em Neossolo Quartzarênico. Revista Brasileira de Zootecnia. 2007;**36**:1715-1728

[13] Jank L, Chiari L, Valle CB, Simeão RM. Forage Breeding and Biotechnology. 1st ed. Campo Grande: Embrapa Gado de Corte; 2013 280 p

[14] Bruulsema T, Lemunyon J, Herz B. Know your fertilizer rights. Crops and Soils. 2009;**42**:13-18

[15] Marschner P, Rengel Z. Nutrient Cycling in Terrestrial Ecosystems. Verlag Berlim: Spring; 2007 291 p

[16] Nesper M, Bünemann EK, Fonte SJ, Rao IM, Velásquez JE, Ramirez B, Hegglin D, Frossard E, Oberson A. Pasture degradation decreases organic P content of tropical soils due to soil structural decline. Geoderma. 2015;**258**:123-133

[17] Alves BJR, Urquiaga S, Jantalia CP, Boddey R. Dinâmica do carbono em solos sob pastagens. In: Santos GA, Silva LS, Canellas LP, Camargo FAO, editors. Fundamentos da matéria orgânica do solo: ecossistemas tropicais e subtropicais. 2nd ed. Porto Alegre: Metropole; 2008 654 p

[18] Fontes SJ, Nesper M, Hegglin D, Velásquez JE, Ramirez B, Rao IM, Bernasconi SM, Bünemann EK, Frossard E, Oberson A. Pasture degradation impacts soil phosphorus

storage via changes to aggregate-associated soil organic matter in highly weathered tropical soils. Soil Biology & Biochemistry. 2014;**68**:150-157

[19] Fernandes LA, Furtini Neto AE, Curi N, Lima JM, Guedes GAA. Fósforo e atividade de fosfatase em dois solos sob diferentes condições de uso. Pesquisa Agropecuária Brasileira. 1998;**33**:1159-1170

[20] Rheinheimer D, Gatiboni LC, Kaminski J. Fatores que afetam a disponibilidade do fósforo e o manejo da adubação fosfatada em solos sob sistema de plantio direto. Ciência Rural. 2008;**38**:576-586

[21] Nash DM, Haygarth PM, Turner BL, Condron LM, McDowell RW, Richardson AE, Watkins M, Heaven MW. Using organic phosphorus to sustain pasture productivity: A perspective. Geoderma. 2014;**222**:11-19

[22] Eberhardt DN, Vendrame PRS, Becquer T, Guimarães MF. Influência da granulometria e da mineralogia sobre a retenção do fósforo em latossolos sob pastagens no cerrado. Revista Brasileira de Ciência do Solo. 2008;**32**:1009-1016

[23] Motta PEF, Curi N, Siqueira JO, Van Raij B, Furtine Neto AE, Lima JM. Adsorção e formas de fósforo em Latossolos: Influência da mineralogia e histórico de uso. Revista Brasileira de Ciência do Solo. 2002;**26**:349-359

[24] Reatto A, Bruand A, Silva EM, Martins ES, Brossard M. Hydraulic properties of the diagnostic horizon of Latosols of a regional toposequence across the Brazilian central plateau. Geoderma. 2007;**139**:251-259

[25] Ker JC. Latossolos do Brasil: uma revisão. Geonomos. 1997;**5**:17-40

[26] Tollefson J. Brazil: The global farm. Nature: News Feature Food. 2010;**466**:554-556

[27] Parfitt RL. The availability pf P from phosphate-goethite brindging complexes, desorption and uptake by ryegrass. Plant and Soil. 1979;**53**:55-65

[28] Raij B. Fertilidade do solo e adubação. Ceres, Potafos: Piracicaba; 1991 343 p

[29] Malavolta E. Adubos e adubação fosfatada. Fertilizantes Mitsui: São Paulo; 1985 61 p

[30] Santos DR, Gatiboni LC, Kaminski J. Fatores que afetam a disponibilidade do fósforo e o manejo da adubação fosfatada em solos sob sistema plantio direto. Ciência Rural. 2008;**38**:576-586

[31] Santos JZL, Furtini Neto AE, Resende AV, Curi N, Carneiro LF, Costa SEVGA. Frações de fósforo em solo adubado com fosfatos em diferentes modos de aplicação e cultivado com milho. Revista Brasileira de Ciência do Solo. 2008;**32**:705-714. DOI: 10.1590/S0100-06832008000200025

[32] Chen CR, Condron LM, Davis MR, Sherlock RR. Effects of plant species on microbial biomass phosphorus and phosphatase activity in a range of grassland soils. Biology and Fertility of Soils. 2004;**40**:313-322

[33] Acosta-Martínez V, Tabatabai M. Enzyme activities in a limed agricultural soil. Biology and Fertility of Soils. 2000;**31**:85-91. DOI: 10.1007/s003740050628

[34] Bais HP, Park S-W, Weir TL, Callaway RM, Vivanco JM. How plants communicate using the underground information superhighway. Trends in Plant Science. 2004;**9**:26-32

[35] Uren NC. Types, amounts, and possible functions of compounds released into the rhizosphere by soil-grown plants. In: Pinton R, Varanini Z, Nannipieri P, editors. The Rhizosphere. New York: Dekker; 2001. p. 19-40

[36] Heuck C, Weig A, Spohn M. Soil microbial biomass C:N:P stoichiometry and microbial use of organic phosphorus. Soil Biology and Biochemistry. 2015;**85**:119-129

[37] Bünemann EK, Oberson A, Liebisch F, Keller F, Annaheim KE, Huguenin-Elie O, Frossard E. Rapid microbial phosphorus immobilization dominates gross phosphorus fluxes in a grassland soil with low inorganic phosphorus availability. Soil Biology and Biochemistry. 2012;**51**:84-95

[38] Groffman PM, Fisk MC. Phosphate additions have no effect on microbial biomass and activity in a northern hardwood forest. Soil Biology and Biochemistry. 2011;**43**:2441-2449

[39] Naples B, Fisk M. Belowground insights into nutrient limitation in northern hardwood forests. Biogeochemistry. 2010;**97**:109-121

[40] Pavinato PS, Dao T, Rosolem CA. Tillage and phosphorus management effects on enzyme-labile bioactive phosphorus availability in Cerrado Oxisols. Geoderma. 2010;**156**:207-215

[41] Gatiboni LC, Kaminski J, Rheinhemer DS, Flores AFC. Biodisponibilidade de formas de fósforo acumuladas em solos sob sistema plantio direto. Revista Brasileira de Ciência do Solo. 2007;**31**:691-699

[42] Gatiboni LC, Brunetto G, Rheinhemer DS, Kaminski J. Fracionamento químico das formas de fósforo do solo: usos e limitações. In: Araújo AP, Alves BJR, editors. Tópicos Em ciência Do Solo. 8th ed. Viçosa: Sociedade Brasileira de Ciência do Solo; 2013. p. 141-187

[43] Daroub SH, Pierce FJ, Ellis BG. Phosphorus fractions and fateof phosphorus-33 in soils under plowing and no-tillage. Soil Science Society of America Journal. 2000;**64**:170-176. DOI: 10.2136/sssaj2000.641170x

[44] Tiecher T, Santos DR, Kaminski J, Calegari A. Forms ofiorganica phosphorus in soil under different long term soil tillage systems and winter crops. Revista Brasileira de Ciência do Solo. 2012;**36**:271-281. DOI: 10.1590QS0100-06832012000100028.

[45] Cunha TJF, Madari BE, Benites V, De M, Canellas LP, Novotny EH, Moutta R, De O, Trompowsky PM, Santos GA. Fracionamento químico da matéria orgânica e características de ácidos húmicos de solos com horizonte a antrópico da Amazônia (Terra Preta). Acta Amazonica. 2007;**37**:91-98. DOI: 10.1590/ S0044 59672007000100010

[46] Zaia FC, Gama-Rodrigues AC, Gama-Rodrigues EF. Formas de fósforo no solo sob leguminosas florestais, floresta secundária e pastagem no Norte Fluminense. Revista Brasileira de Ciência do Solo. 2008;**32**:1191-1197

[47] Rosset JS, Guareschi RF, Pinto LASR, Pereira MG, Lana MDC. Phosphorus fractions and correlation with soil attributes Ina a chronosequence of agricultural under no-tillage. Semina: Ciências Agrárias. 2016;**37**:3915-3926. DOI: 10.5433/1679-0359

[48] Dieter D, Elsenbeer H, Turner BL. Phosphorus fractionation in lowland tropical rainforest soils in central Panama. Catena. 2010;**82**:118-125

[49] Cross AF, WHA S. Literature review and evaluation of the Hedley fractionation: Aplications to the biogeochemical cycle of soil phosphorus in natural ecosystems. Geoderma. 1995;**64**:197-214

[50] Novais RF, Smyth TJ, Nunes FN. Fósforo. In: Novais RF, Alvarez VVH, Barros NF, Fontes RLF, Cantarutti RB, Neves JCL, editors. Fertilidade Do Solo. 1st ed. Viçosa: Sociedade Brasileira de Ciência do Solo; 2007. p. 471-537

[51] Novais RF, Smyth TJ. Fósforo em solo e planta em condições tropicais. Viçosa, MG: Universidade Federal de Viçosa; 1999 399 p

[52] Silva FC, Raij BV. Disponibilidade de fósforo em solos avaliada por diferentes extratores. Pesquisa Agropecuária Brasileira. 1999;**34**:267-288

[53] Condron LM, Newman S. Revisiting the fundamentals of phosphorus fractionation of sediments and soils. Journal of Soils and Sediments. 2011;**11**:830-840

[54] Tokura AM, Furtini Neto AE, Carneiro LF, Curi N, Santos JZL, Alovisi AA. Dinâmica das formas de fósforo em solos de textura e mineralogia contrastantes cultivados com arroz. Acta Scientiarum. Agronomy. 2011;**33**:171-179. DOI: 10.4025/actasciagron.v33i1.1435

[55] Alovisi AMT, Furtini Neto AE, Serra AP, Alovisi AA, Tokura LK, Lourente ERP, Silva RS, Silva CFB, Fernandes JS. Phosphorus and silicon fertilizer rates effects on dynamics of soil phosphorus fractions in oxisol under common bean cultivation. African Journal of Agricultural Research. 2016;**11**:2697-2707. DOI: 10.5897/AJAR2016.11304

[56] Leite JNF, Cruz MCP, Ferreira ME, Andrioli I, Braos LB. Frações orgânicas e inorgânicas do fósforo no solo influenciadas por plantas de cobertura e adubação nitrogenada. Pesquisa Agropecuária Brasileira. 1889;**51**:1880-1889. DOI: 10.1590/S0100-204X2016001100010.

[57] Hedley MJ, Stewart JWB, Chauhan BS. Changes in inorganic and organic soil phosphorus fractions induced by cultivation practices and by laboratory incubations. Soil Science Society of American Journal. 1982;**46**:970-976. DOI: 10.2136/sssaj1982.03615995 004600050017x.

[58] Cordell D, Drangert J-O, White S. The story of phosphorus: Global food security and food for thought. Global Environmental Change. 2009;**19**:292-305

[59] Camacho MA, Silveira LPO, Silveira MV. Eficiência de genótipos de Brachiaria brizanthaStapf. (Syn: *Urochloa brizantha*) na produção de biomassa sob aplicação de fósforo. Arquivo Brasileiro de Medicina Veterinária e Zootecnia. 2015;**67**:1133-1140

[60] Calderón LS. Genetic characterization of morpho-physiological responses of the root system of *Arabidopsis thaliana* (L.) Heynh., for phosphorus deficiency [thesis]. Ciudad de México: Centro de Investigación y de Estudios Avanzados del Instituto Politécnico Nacional – CINVESTAV; 2006

[61] Procópio SO, Santos JB, Pires FR, Silva AA, Mendonça ES. Absorption and utilization of phosphorus by soybean and bean and weeds. Revista Brasileira de Ciências do Solo. 2005;**29**:911-921

[62] Lajtha K, Harrison AF. Strategies of phosphorus acquisition and conservation by plant species and communities. In: Tiessen H, editor. Phosphorus in the Global Environmental: Transfers, Cycles and Management. 1st ed. Chichester, UK: Wiley; 1995. p. 139-147

[63] Schenk MK, Barber SA. Root characteristics of corn geno-types as related to P uptake. Agronomy Journal. 1979;**71**:921-924

[64] Kunze A, Costa MD, Epping J, Loffaguen JC, Schuh R, Lovato PE. Phosphatase activity in sandy soil influenced by mycorrhizal and non-mycorrhizal cover crops. Revista Brasileira de Ciência do Solo. 2011;**35**:705-711

[65] Bayon RCL, Weisskopf L, Martinoia E, Jansa J, Frossard E, Keller F, Föllmi KB, Gobat JM. Soil phosphorus uptake by continuously cropped *Lupinus albus*: A new microcosm design. Plant and Soil. 2006;**283**:309-321

[66] Dalla MC, Lovato PE. Fosfatases na dinâmica do fósforo do solo sob culturas de cobertura com espécies micorrízicas e não micorrízicas. Pesquisa Agropecuária Brasileira. 2004;**39**:603-605

[67] Nahas E. Microrganismos do solo produtores de fosfatases em diferentes sistemas agrícolas. Bragantia. 2002;**61**:267-275

[68] Turner BL, Haygarth PM. Phosphatase activity in temperate pasture soils: Potential regulation of labile organic phosphorus turnover by phosphodiesterase activity. The Science of the total environment. 2005;**344**:27-36

[69] Verbruggen E, Van Der Heijden MGA, Weedon JT, Kowalchuk GA, Röling WFM. Community assembly, species richness and nestedness of arbuscular mycorrhizal fungi in agricultural soils. Molecular Ecology. 2012;**21**:2341-2353

[70] Johnson NC. Resource stoichiometry elucidates the structure and function of arbuscular mycorrhizas across scales. New Phytologist. 2010;**185**:631-647

[71] Simpson RJ, Richardson AE, Nichols SN, Crush JR. Pasture plants and soil fertility management to improve the efficiency of phosphorus fertiliser use in temperate grassland systems. Crop and Pasture Science. 2014;**65**:556-575

[72] Soares WV, Lobato E, Sousa DMG, Rein TA. Avaliação de fosfato natural de Gafsa para recuperação de pastagem degradada em Latossolo Vermelho-Escuro. Brazilian Journal of Agricultural Research. 2000;**35**:819-825

[73] Moreira A, Fageria NK, Souza GB, Freitas AR. Production, nutritional status and chemical properties of soils with addition of cattle manure, reactive natural phosphate and biotite schist in Massai cultivar. Revista Brasileira de Zootecnia. 2010;**39**:1883-1888

[74] Ramos SJ, Faquin V, Rodrigues CR, Silva CA, Boldrin PF. Biomass production and phosphorus use of forage grasses fertilized with two phosphorus sources. Revista Brasileira de Ciência do Solo. 2009;**33**:335-343

[75] Ministry of Agriculture, Livestock and Food Supply (MAPA), Normative Instruction No. 46 November 22, 2016. Available from: http://www.agricultura.gov.br/assuntos/insumos-agropecuarios/insumos-agricolas/fertilizantes/legislacoes

[76] Martha JGB, Vilela L, Sousa DMG, Barcellos AO. Adubação Nitrogenada. In: JGB M, Vilela L, DMG S, editors. Cerrado: Uso Eficiente de Corretivos e Fertilizantes Em Pastagens. 1st ed. Embrapa Cerrados: Planaltina; 2007. p. 117-144

[77] Ovalles FA, Collins ME. Variability of northwest Florida soils by principal component analysis. Soil Science Society of America Journal. 1988;**52**:430-1435. DOI: 10.2136/sssaj1988.03615995005200050042x

[78] Yang Z, Culvenor RA, Haling RE, Stefanski A, Ryan MH, Sandral GA, Kidd DR, Lambers H, Simpson RJ. Variation in root traits associated with nutrient foraging among temperate pasture legumes and grasses. Grass and Forage Science. 2015;**72**:93-103. DOI: 10.1111/gfs.12199

Chapter 7

Best Management Practices (BMPs) for Nitrogen Fertilizer in Forage Grasses

Ademar Pereira Serra, Marlene Estevão Marchetti,
Elisângela Dupas, Simone Candido Ensinas,
Elaine Reis Pinheiro Lourente,
Eulene Francisco da Silva,
Roberto Giolo de Almeida,
Carla Eloize Carducci and
Alessandra Mayumi Tokura Alovisi

Additional information is available at the end of the chapter

http://dx.doi.org/10.5772/intechopen.70345

Abstract

There is a concern about the growing population and limitation in natural resources which are taking the population to direct its agricultural systems into a more productive and efficient activity, looking to avoid a negative impact on the surrounding environment. The industry energy expended to produce nitrogen (N)-fertilizer is considered an indirect consumption of energy in agriculture, which is higher with an increasing forage yield. Nitrogen is the key nutrient associated with high-yielding production in forage grass and grain crops. The aim of this chapter is to introduce the best management practices (BMPs) for N-fertilizer application in forage grasses to improve N-use efficiency, since the most economical way to feed livestock is forage plants where its potential biomass production is not well explored. The BMPs basically follow three management practices: (1) soil nutrient availability and forage requirement, (2) fertilizer application, and (3) decrease in nutrient losses from soil. In order to take a decision on applying N-fertilizer to accomplish forage grasses production with social, economic, and environmental benefits, the N-fertilizer use in forage grasses is going to follow the "Right rate, Right source, Right place, and Right time (4R) nutrient stewardship." The application of the 4R's nutrients stewardship is directly associated with economic, social, and environmental impact. The capacity of the 4R's implementation worldwide turns into a best guide to improve the striving of better N-use efficiency in forage grass. The 4R's are interrelated; thus, the recommendation of N-fertilizer rates cannot be prescribed without the combination of the 4R's where a whole system to be followed should be considered to decide about N-fertilizer in pasture. Consequently, any decision in one of the 4R's is going to affect the expected N-fertilizer results and dry matter production.

Keywords: soil fertility, N-fertilizer recovery, N-use efficiency, *Brachiaria* spp., *Panicum maximum*

1. Introduction

A remarkable increase in cattle herd in Brazil occurred from 1977 to 2012, with the Midwest region showing a high increase [1]. However, extending to new frontiers for Brazilian cattle herd is constraint to limited areas and consequently the yield of forage grass needs to be increased. Thus, the intensification of livestock production tends to intensive management. Increasing livestock production due to the world demand for meat is quite associated with an improvement in forage production. To feed grazed animal in livestock, tropical forages are the cheapest source of food in Brazil. *Brachiaria* spp., *Panicum maximum*, and *Stylosanthes* spp. are the majorly cultivated forage in Brazil. Among the gender *Brachiaria* spp., *B. brizantha* cv. Marandu is majorly cultivated in Brazil with 45% of the whole area with cultivated pastures [2], which correspond to approximately 45 million hectares. However, in Brazil there are millions of hectares with degraded pastures, due to soil with low fertility and no replacement of nutrients removed from soil over the years.

The uses of synthetic fertilizers in agriculture are primordial to sustain the growing population worldwide, which tends to increase linearly with population growth at least until 2050 [3]. Different from phosphorus (P), potassium (K), calcium (Ca), magnesium (Mg), and micronutrients [boron (B), zinc (Zn), iron (Fe), manganese (Mn), and copper (Cu)], N needs to be applied to every cropping season, because N does not remain longer in soil profile. Soil organic matter (SOM) is the main source of N, which can compile 95% of the total N and 5% remain as NO_3-N and NH_4-N. However, the amount of N in soil varies a lot due to many factors including soil clay content, moisture, aeration, temperature, tillage, rainfall, and so on, which implies a high N dynamic in soil and consequently the problem in measuring the inorganic soil N-content (NO_3-N and NH_4-N).

Nitrogen and phosphorus represent the most limited macronutrients in tropical forages. Forage grasses are quite responsive to N-fertilizer; its applications may result in increasing crude protein (CP), number of tillers and leaves, and consequently dry matter (DM) production, since other production factors are not limited [4]. Nitrogen-fertilizer in forage grasses has shown low-use efficiency on tropical climate region. Nitrogen-fertilizer is usually applied in intensive management systems and results in high NH_3-N losses, causing low N-use efficiency (NUE).

Using N-fertilizer in forage grass must be well thought because overrates of N-fertilizer can promote excess of forage biomass above the capacity of consumption of animal grazing; thus the stocking rate needs to be adjusted in accordance with forage availability. We must keep in mind that the response to N-fertilizer is closely related to the adequate content of P, K, and other nutrients available in soil. Just N application in forages cannot result in a satisfactory increase of biomass production, if there are constraints of other nutrients in soil, occasioning in low NUE [5].

In order to increase N-fertilizer efficiency in forage grasses, one must keep in mind that the concepts of best management practices (BMPs) for fertilizer application must be followed, which combines the **R**ight source, **R**ight rate, **R**ight place, and **R**ight time's (4R) nutrient stewardship [5]. These concepts are universally applied, but the adjustment of the 4R's recommendation depends on site-specific conditions, thus the use of a specific nutrient rate for one location is unpractical and even more unrealistic for other regions. Based on the earlier explanation, this chapter introduces the concept of the 4R's to direct BMPs to improve NUE in forage grasses, in order to combine all practices forward the best economic, social, and environment conditions.

2. Nitrogen-use efficiency

The concept of best management practices (BMPs) for fertilizer use is quite important to improve N-use efficiency (NUE) in forage grasses. In forage breeding programs, the application of NUE could be a useful tool to select forage genotypes with higher capacity of biomass production under lower amount of N-fertilizer rate. The improvement of NUE can result in higher quantity and quality of dry matter production on grassland. Even with the majority of methodologies conducted using grain crops as shown in **Table 1**, the adaptation of these procedures can be well applied in forage grasses.

In general, the concept of NUE is associated with higher yield and low N-fertilizer input; thus, it can be achieved if other N-management has already applied. The use of N-sources can change NUE and even the rate, place, and time of N-fertilizer. To improve NUE, it is a complex combination, where the concepts of BMPs must be followed step by step, which are associated with the Right rate, Right source, Right place, and Right time (4R's) of N-fertilizer application on forage grass or grain crops [5]. Thus, all the concepts reported in the following subheadings will achieve a higher dry matter production with low N-input in the forage production, combined with social, economic, and environment benefits.

N-use efficiency index	Calculation	References
AEN = Agronomic efficiency of applied N (kg yield increased per kg N applied)	AEN = (YN–Y0)/FN	[51, 52]
NER = N-efficiency ratio (kg of yield per kg of N in tissue)	NER = [(Units of yield)/(Units of N in tissue)]	[53]
PEN = Physiological efficiency of applied N (kg yield increase per kg increase in N-uptake from fertilizer)	PEN = (YN–Y0)/(UN–U0)	[51, 52]
NFR = N-fertilizer recovery (%)	NFR (%) = [(UN–U0)/(FN)].100	[52, 54]
SNS = Soil N-recovery (%)	GNCu/GNCf multiplied by 100	[52, 54]

FN, the amount of (fertilizer) N applied (kg ha^{-1}); YN, forage yield with applied N (kg ha^{-1}); Y0, crop yield (kg ha^{-1}) in a control treatment with no N; UN, total plant N-uptake in aboveground biomass at maturity (kg ha^{-1}) in a plot that received N; U0, the total N-uptake in aboveground biomass at maturity (kg ha^{-1}) in a plot that received no N.

Table 1. Agronomic indices for N-use efficiency for forages.

Another manner to assess fertilizer application is through the bioeconomic efficiency that is compiled by the conversion efficiency of N-fertilizer into forage dry matter produced, by the efficiency that the produced forages are consumed by the grazing animal, and through the efficiency to convert forage into animal products [6].

3. The 4R's for N-fertilizer in forage grass

The consequences of uncorrected N-management practice can have an impact on the increasing greenhouse gas (GHG) emission, NH_3-N volatilization, N-runoff, and water eutrophication, which are directly associated with a negative impact on environment. The success of N-fertilizer in forage depends on the combination of rate, source, place, and time of application.

In general, BMPs are followed by three management practices that include the combination of soil nutrient availability and forage requirement, fertilizer application, and decreased N-losses from soil. The BMPs for fertilizer application are based on the 4R's nutrient stewardship (Right source, Right rate, Right place and Right time) (**Figure 1**). These concepts combined direct understanding on how one can advance under a sustainable agriculture. The 4R's are considered universal; then the scientific practices that direct the 4R's can be applied and adjusted in site-specific around the world. Therefore, for each region and even each farmer there will be a set of practices that are site-specific to implement the 4R's nutrient stewardship. The idea to have just one common recommendation for N-fertilizer cannot be followed any longer.

All strives to implement the 4R's nutrient stewardship in grassland must be incentivized to improve NUE. The concern about high production cost and surrounding environment impact is quite evident in agricultural system, and N-fertilizer has a great quota of increasing these problems due to its high mobility in soil through NO_3-N leaching, ammonium volatilization (NH_3-N), and nitrous oxide (N_2O) emission [7].

In order to apply the 4R's nutrients stewardship, it is important to keep in mind that the 4R's concept is directly associated with economic, social, and environmental impact (**Figure 1**).

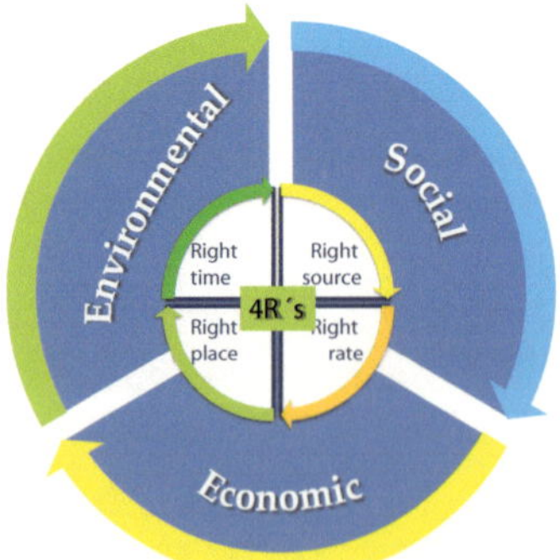

Figure 1. The 4R's nutrient stewardship for N-fertilizer in pastures. Adapted from Bruulsema et al. [5] and elaborated by the authors.

http://dx.doi.org/10.5772/intechopen.70345

The concepts shown in **Figure 1** are interlinked; therefore, any decision in one of the 4R's can influence directly on the results of N-fertilizer management and consequently on NUE.

3.1. Right rate

Nitrogen requirement for tropical forage and soil N-supply must be balanced with N-fertilizer rates, which is the most important nutrient removed from the soil in perennial pasture. Quantification of soil N-availability is quite difficult; because N is very dynamic, changing from organic to inorganic forms depending on weather conditions, and soil feature, as aeration, bulk density, soil water content, moisture, etc. Thus, N-quantification through soil analysis becomes incorrect or not much realistic. On the other hand, for other nutrients like P, K, Ca, Mg, S, and micronutrients, soil nutrient quantification using chemical extractors can result in accurate diagnosis.

Preview researches related to total N in soil quantify that 95% of the whole N in soil are combined in organic compounds that is available for plant uptake after the mineralization process, resulting in inorganic forms of N (NO_3-N and NH_4-N) [8]. The total N in soil with pasture can achieve on average 2 Mg ha^{-1} of organic N with 5 years of pasture implemented under rotation with soybean and maize [9]. Consequently, the total N in soil and labile N are indicative of soil supply (**Figure 2**). In comparison to other crops and intercropping, *B. ruziziensis* and *B. brizantha* cv. Marandu cultivated without intercropping showed higher labile N in soil [9]. Thus, even with constraints in evaluating soil N-supply, the quantification of SOM, total N, and labile N can direct some information to guide the recommendation of N-fertilizer rates.

Nitrogen is required by tropical forage in high amounts, and the answer has varied among the forage species which range from 200 to 1.800 kg N ha^{-1} per year [9]. In order to recommend N-fertilizer rates, besides the factors already mentioned, it is necessary to take into account

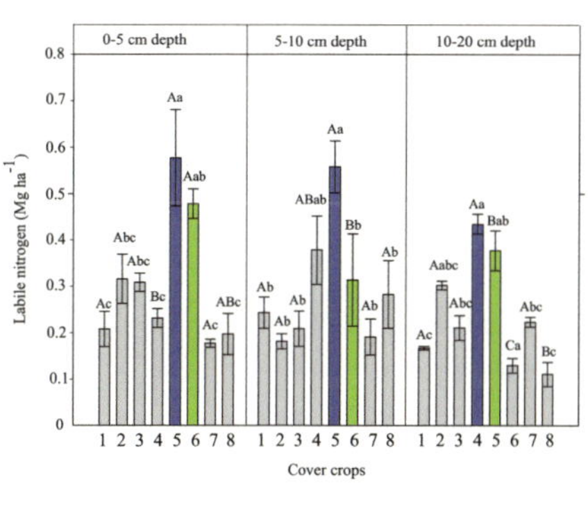

Figure 2. Cover crops and soil profile effects on labile nitrogen. Different uppercase letters indicate significant difference ($p \leq 0.05$) among different soil depths within the same cover crop while the different lowercase letters indicate significant difference ($p \leq 0.05$) among different cover crops within the same soil depth by Tukey test of means. The error bars are the standard errors. (1) Fall–winter corn; (2) Intercropping fall–winter corn with *B. ruziziensis*; (3) Intercropping fall–winter corn with *B. brizantha* cv. Marandu; (4) Intercropping fall–winter corn with *C. spectabilis*; (5) *B. ruziziensis*; (6) *B. brizantha* cv. Marandu; (7) *Pennisetum glaucum* L; (8) Fallow area. Fonte: Ensinas et al. [9].

the capacity of N-fertilizer use efficiency by forage plants and its impact on the stocking rate. Thus, the amount of cattle herd is capable to consume the forage and the management for feeding to avoid excess and absence of forage [4].

Even with high response to N-fertilizer rate by tropical forage, resulting in increasing crude protein (CP), and other bromatological features [10], the risk of higher downward residual NO_3-N movement in soil and high concentration of NO_3-N in biomass due to N-fertilizer above plant requirement must be considered to avoid environment contamination and animal health problems [11, 12].

In order to increase forage biomass production, the balance of nutrients in soil and plants must be considered, and the constraint of other nutrients cannot be replaced by an increasing rate of N-fertilizer, and thus the expected dry matter production associated with plant requirement is the guide to determine the appropriate N-fertilizer rate. The low S supply may increase NO_3-N accumulation and soluble protein in plant tissue [13], due to possible restriction in NO_3-N reductase enzyme [14].

In a condition without N-fertilizer, biological fixation of $N_{2,}$ and atmospheric deposition, the N available for forage plant uptakes is just the N mineralized from SOM [8]. Thus, avoiding N-fertilizer in forage can deplete its biomass production over time, since soil N-mineralization is slow and not able to supply forage requirement for a long time. One must keep in mind that grazed animals must consume the biomass production in forage; therefore, the stocking rate management must be taken into consideration to decide the N-fertilizer rate.

Usually, the recommendation of N-fertilizer in forage follows the concept of yield response with an expected production and stocking rate associated to consume the excess production, both associated with N-fertilizer rate experiments in site-specific conditions.

3.2. Right source

Soil organic matter (SOM) is the highest supplier of N for forage growth, which can reach 85% of the whole N required by forages [15]; thus, the absence of N-fertilizer rates to replace the plant uptakes can conduct to pasture degradation. Forage plants uptake N majority through nitrate (NO_3-N) and ammonium (NH_4-N) [16], which depend on their contents in soil solution (**Figure 3**). The proportion of NO_3-N and NH_4-N is determined by soil conditions, NO_3-N can be predominant in aerobic condition where nitrification can occur. On the other hand, NH_4-N can be predominant in acid soil and anaerobic soil. However, the assimilation of NO_3-N in plants tends to expend to be more energetic than NH_4-N assimilation. Through the action of inducible enzyme (nitrate reductase), the NO_3-N is reduced to NH_4-N and finally incorporated to glutamine [16]. When the uptake occurs through NH_4-N, the expenditure of energy with the NO_3-N reduction is solved.

Good result in forage production was shown when NO_3-N and NH_4-N had 70 and 30% in soil, respectively [17, 18]. On the other hand, the use of NO_3-N:NH_4-N mixture at the ratio of 55:45% instead of the NO_3-N solely as N source in the nutrient solution enhanced the production of tillers (30%), leaves (20%), and the leaf area surface (30%) of *P. maximum* Jacq. cv. Aruana (*Aruana guineagrass*) [12].

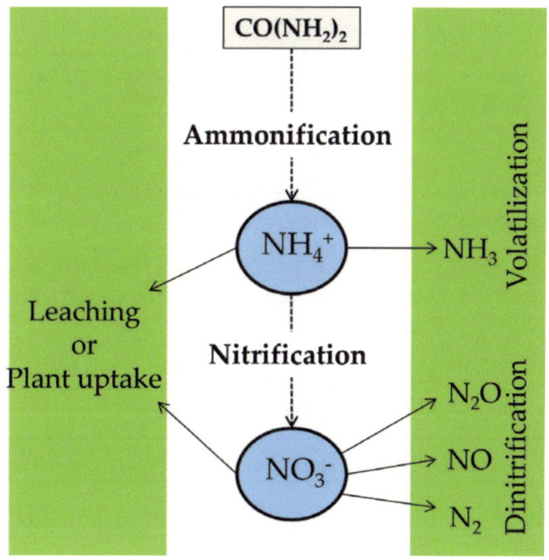

Figure 3. Summary of urea reaction on soil. Elaborated by authors.

Anyway, both forms of N (NO_3 or NH_4) are incorporated to amino acids through the glutamine synthetase-glutamate [19]. However, the major problem is the accumulation of NO_3-N in forage, because high NO_3-N concentrations (above 4500 mg kg^{-1} of dry matter considered highly toxic) can cause animal mortalities [20].

There are many N-fertilizer sources for use in forage (**Table 2**). Nitrogen-fertilizer sources with a higher content of N are preferable to apply in grassland due to lower transport cost and facility of application. Urea, ammonium sulfate, and nitrate are the three majority sources of N used in forage grass. However, each of them has peculiarity in relation to using in forage with its pros and cons.

On the one hand, urea [$CO(NH_2)_2$] shows the highest concentration of N (**Table 2**), and lower price per N unit in its composition, and causes less soil acidification compared with ammonium sulfate [21], on the other hand, urea is the N-fertilizer source that causes higher NH_3-N losses through volatilization [16], and consequently lower NUE.

In order to decrease losses of NH_3-N from urea source, the application in season with higher rainfall would be indicated, and as other alternative, the use of urea with the inhibitor of urease can be a viable alternative in forage grass; nevertheless, the still high cost of urea with urease inhibitor can restraint its use in forage grass. In studying with six sources of N-fertilizer *B. brizantha* cv. Marandu, it was observed that N-sources do not change the performance of dry matter; however, urea with urease inhibitor and polymer-coated urea improved NUE [22]. Urea mechanical incorporation or rain after N-fertilizer application can decrease the amount of NH_3-N losses, due to higher speed of urea incorporated in soil [11].

Ammonium sulfate [$(NH_4)_2SO_4$] has 22% of sulfur (S) and 20% of N in composition (**Table 2**); this S content is quite interesting because of the improvement of N-use efficiency by forages.

Fertilizer source	Minimum guarantee	
	Nutrient content and form	Nutrient solubility/granulometry
Urea	45% of N	Total content of N.
Ammonium sulfate	20% of N 22% of S	Total content of N and S.
Fosfato Diamônico (DAP)	17% of N 45% of P_2O_5	Total content of N and P_2O_5 content soluble in CNA plus water and minimum of 44% soluble in water.
Fosfato Monoamônico (MAP)	9% of N 48% of P_2O_5	Total content of N and P_2O_5 content soluble in CNA plus water and minimum of 44% soluble in water.
Ammonium nitrate	32% of N	Total content of N.

Fonte: Adapted from Agricultural Ministry (MAPA), Normative Instruction No. 46, November 22, 2016.

Table 2. Major N-sources, specification of the simple solid sources of nitrogen with minimum granulometric guarantee.

Nevertheless, ammonium sulfate acidifies the soil due to the nitrification process. Among urea, ammonium sulfate, and potassium nitrate fertilizers applied in forage grasses, ammonium sulfate has the highest capacity to increase soil acidity [23].

Nitrogen and S are closely related to plant metabolism, resulting in increasing protein content when both are in adequate balance for forage. The ratio of N/S for *B. brizantha* (Hochst. ex A. Rich.) Stapf. cv. Marandu was 10/1; the fertilization promoted high yield, adequate N- and S-concentrations for plant metabolism, and forage production, as well as maintained and/or raised the soil fertility in relation to these nutrients [24].

In a study published by [25], the optimum N/S ratio was 14.02 in a high-yielding population defined through method DRIS (diagnosis and recommendation-integrated system). In order to use ammonium sulfate source, it is recommended to apply liming to correct the acidity promoted by this N-source.

Another way to improve N-availability to forage grass is the mixture of legumes and pasture. In Brazil, the native *Stylosanthes* spp. are used in mixture with *Brachiaria* spp. The use of a mixture of legume with gramineas is quite interesting to reduce N-fertilizer rate due to legume capacity to fix molecular N_2 from atmosphere, but it is still a challenge to maintain the right proportion of these two species without competition, because the legume tends to disappear due to competition [26, 27].

Besides legumes, which has symbiotic association with microorganism in root system, *Brachiaria* spp. may be partially N supplied by inoculation with *Azospirillum amazonense*. The association of *B. brizantha, B. decumbens,* and *B. humidicola* with *A. amazonense* was observed by [28], which reported the capacity of *A. amazonense* in producing hormones as indole-3-acetic acid (IAA, 3-IAA). Accordingly, the inoculation of *A. amazonenses* with *Brachiaria* spp. may result in positive interaction, resulting in such improvement in promoting the growth of forages. The association of *Brachiaria* spp. and *Ancylostoma brazilienses* may promote increasing

N-accumulation respective to 40 kg ha^{-1} of N-fertilizer rate. Besides previous information about *Brachiaria* spp. inoculation with *Azospirillum* spp., much more researches must be taken for better understanding about the efficiency of *Brachiaria* spp. inoculation.

Microorganisms in soil provide great contribution in N-availability through mineralization of SOM, basically due to enzyme activity in mineralization process. There are important correlations between microbial biomass carbon (C-MBC) and microbial biomass nitrogen (N-MBN) [29]. In general, in crop rotation system the rate of C-MBC:C-organic and N-MBM:N-total is 1.1 and 2.6%, respectively, while in single crop the rate is 0.8 and 2.1% [30]. As reported by [29], in pasture system values of 2.4 and 3.2% (N-MBM:C-MBC) were observed.

Under pasture system, the species of forages and soil cover promoted significant effect on C-MBC and enzymatic activity; therefore, these enzymes can be used as indicator of soil quality [29]. Even with equal C-MBC and N-MBN, the activity of urease, protease, and dehydrogenase in pasture was different [29]. The intensity and forms of different plant species influenced in N-cycle require further studying. As reported by [31], in order to increase the availability of inorganic N through the action of root exudates, the N-microbial mineralization depends on C availability and labile N. Depending on the type of exudation, the losses of N_2 to atmosphere can be increased due to reduction of N_2O to N_2 [31]. The importance of microorganism in improving N through mineralization is crucial in soil, which needs more researches in forage grass to improve our knowledge in N dynamic into this production system.

3.3. Right place

Broadcasting without incorporation is a common manner to apply N-fertilizer in forage grasses in tropical forage. Even with less effective for improvement of NUE, N-fertilizer broadcasting is considered the most practical procedure to apply in larger areas of livestock in tropical climate as the case of Brazil. Besides, this widely used procedure, there are other placement methods that can be used in forage and have already been applied successfully, as the case of banding N-fertilizer incorporated in soil [32]. The N-fertilizer incorporation is the most effective way to decrease NH_3-N and increase NUE when N-fertilizer source is the common urea. Conversely, the mechanical incorporation of N-fertilizer in pasture sometimes is not possible due to the absence of adequate implement faced by most farmers. Broadcasting ammonium nitrate and ammonium sulfate on soil surface is effective due to low NH_3-N gas volatilization from both N-fertilizer sources.

Applications of urea broadcasting usually show a lower capacity of N-fertilizer recovery, resulting in biomass production of forages below the expected. Urea applied in soil depth markedly reduces the NH_3-N gas volatilization without causing serious damage to forages; however, it is not common among Brazilian farmers. The placement of urea on soil surface can decrease above 40% of the whole N-fertilizer applied in forage grass [15]. Ammonia volatilization resulted from urea application may cause environmental impact, with increasing the NH_3-N in site-specific surrounding of 85% of the total NH_3-N volatilized, and the remaining NH_3-N is conducted to another region through winds [33].

Urea with the inhibitor of urease is an alternative to reduce NH_3-N volatilization and N_2O when urea is broadcasting without incorporation [34], occasioning a lower content of NO_3-N and NH_4-N in soil, but increasing the plant uptake due to higher availability though time [35].

3.4. Right time

In order to optimize N-fertilizer use efficiency for pastures, the right time is decisive to achieve better results in terms of nutrient uptake and biomass production. The time of N-fertilizer application in forage grass varies through the growing season and higher demand of nutrients by plants. In tropical region, N-uptake is highly demanded in summer season, because of the highest biomass accumulation due to higher rainfall, adequate temperature, and sunlight for optimum forage growth (**Figure 4**). The highest plant N-requirement is the right time for N-fertilizer in forage grasses [36], because it tends to improve N-uptake by plants and consequently NUE. For implementation of pasture, the application of N-fertilizer used to show low efficiency due to low plant requirement in the beginning of growth, and SOM mineralization tends to be enough for initial growth.

Even with the suggestion for N-fertilizer time shown in **Figure 4**, time is dependent of intensive or extensive livestock system. In rotated grazing, it is used to apply N-fertilizer right after rotated animal. On the other hand, in extensive system N-fertilizer can follow the application time as suggested in **Figure 4**. The rainfall information needs to be obtained in site-specific

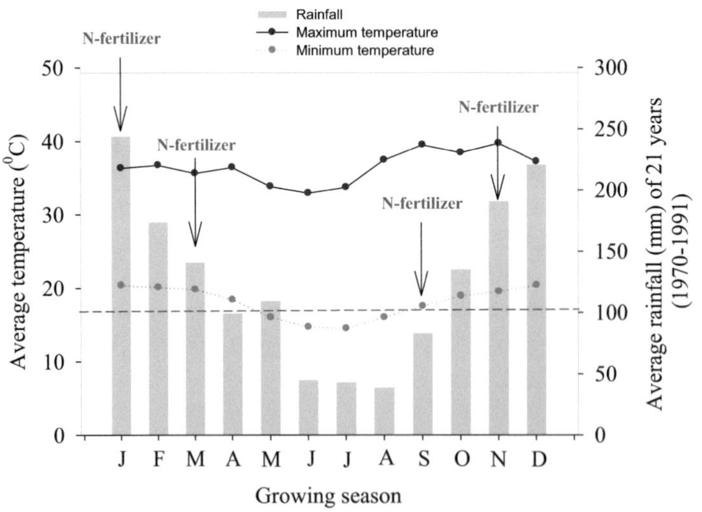

Figure 4. Average rainfall, and maximum and minimum temperature of 21 years. Data from meteorological station of Universidade Federal da Grande Dourados (UFGD), Mato Grosso do Sul State, Brazil. A dashed line means the boundary of water limitation to produce forage dry matter.

region to manage and develop the right time to maximize the NUE by forage grass. N-fertilizer application in dry season tends to show lower NUE due to water limitation for plant nutrient uptake, which is not recommended because of cost-effective being low.

Summer season is considered the best moment to apply N-fertilizer in forage grasses, resulting in higher NUE due to rainfall enough for better growth. In order to avoid the seasonality of forage growth, even with low NUE in comparison to summer season, the application of N-fertilizer at the end of summer season (March) can help to decrease the growth seasonality (**Figure 4**). In a study with rates of N and irrigation for *B. brizantha* cv. Marandu in two periods of the year (dry and wet season) in the IIha Solteira City, Sao Paulo State, Brazil, it was observed that in the wet season, the average DM yield of Marandu grass was 1.9 t ha^{-1} in the irrigated experiment and 1.8 t ha^{-1} in the non-irrigated experiment, that is, with no statistically significant difference [37]. In the dry season, the average DM yields were 8.1 t ha^{-1} in the irrigated treatment and 4.4 t ha^{-1} under non-irrigated treatment. The irrigated experiment produced 55% more than the non-irrigated one.

4. N-concentration and crude protein (CP) content in forage dry matter

The contents of CP are affected by N-fertilizer rates applied in each forage cut or rotated grazing, and by physiologic age of forage [12]. The measurement of CP in dry matter (DM) of forage is directly related to N-concentration in DM, where the amount of CP is the multiplication of N-concentration by 6.25 [8]. The coefficient is related to N-proportion in vegetal protein. However, this content of CP does not reflect just the real amount of crude protein because it is related to all N forms in tissue, even the NO_3-N accounts to result in this amount of CP [10].

Nitrogen balanced in forage can result in higher leaf/stem ratio, palatability, and succulence in forages [10]. According to Cornell or CNCPS model [38], the CP of forage plants and foods is divided into five fractions. The soluble part of protein is divided by A and B1 fractions. The fraction A is the N no-protein (NnP), since this fraction is highly soluble in rumen. The remaining B1 fraction is part of true protein, which also shows fast degradation in rumen. The fraction C corresponds to the unavailable protein and it is the part of protein content in acid detergent fiber (ADF), non-soluble N in acid detergent (nADF). These associations of lignin result in tannin complex and products from Maillard reactions that are resistant to microbial enzymatic degradation. The fraction of N in neutral detergent fiber (NDF) is denominated by N non-soluble in neutral detergent (nNDF). Another form of available protein in plant is the subtraction of nADF and nNDF, which are designated as the fraction B3; however, the rate of degradation is quite slow. The B2 fraction is the last, which shows medium degradation and is considered the non-soluble protein fraction; thus, the B2 fraction does not make a part of the cell wall and non-protein N.

Crude protein fractionation is not commonly done in researches related to tropical forages; however, in some scientific results related to N-fertilizer an increase in CP and a decrease in nADF (fraction C) occur. The decrease in fraction C is desirable, because this fraction compiles the non-soluble protein, which is not degraded in the rumen [39–41].

5. Greenhouse gas (GHG) emissions in grazed pasture under N-fertilizer

The three most important greenhouse gases (GHG) in the atmosphere are carbon dioxide (CO_2), methane (CH_4), and nitrous oxide (N_2O). Both of them are strongly affected by N-fertilizer in agricultural system [42], with a major responsibility of agriculture for methane (CH_4) and nitrous oxide (N_2O) emissions [43, 44].

Nitrous oxide is 310 times more dangerous for stratospheric ozone (O_3) than CO_2 and shows a lifetime of 112 years in the stratosphere [45]. Nitrous oxide emission in grazed pasture is associated with animal stocking rate, animal excreta (urine and dungs), content of soil NO_3-N, N-fertilizer rates, tillage, soil moisture, soil compaction, and other process that affect soil aeration [7, 42].

Nitrous oxide is a way of N-losses in grazed pasture, as well as NO_3-N leaching and NH_3-N volatilization; both forms of N-losses are related to environment depletion [47]. Nitrogen-fertilizer efficiency is directly related to N_2O emission, which is termed denitrification process [16, 46, 47]. In Brazilian Cerrado, the NH_3-N and N_2O through denitrification were observed to be the most important process of N-losses from cattle excreta for 7 months of rainy season in extensive pasture [48]. As reported by [7], the emission of N_2O in grazed pastures is partially associated with C and N deposited from the animal excreta on soil (urine and dungs) under anaerobic conditions, as the case of soil compaction caused by animal trampling.

The anaerobic circumstance can be observed in wet soil after animal trampling [7]; thus, the N-fertilizer applied under wet condition can increase the N_2O emission through NO_3-N content in soil that can be denitrified in pasture. The use of N-fertilizer sources with the inhibitor of urease and nitrification can reduce the N_2O emission [49], due to slow process of urea hydrolase and permanence of NH_4-N form instead of NO_3-N.

Urine and dungs excreta by grazed animals are responsible for a great source of N_2O emission in grazed pastures. Integrated crop-livestock-forestry (ICLFS) or livestock-forestry system (ILFS)

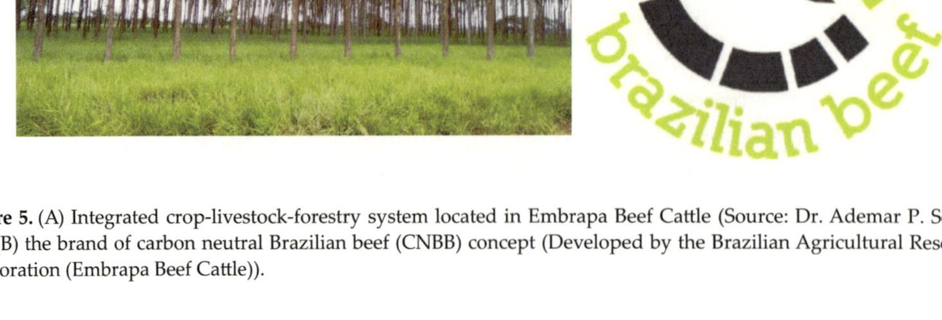

Figure 5. (A) Integrated crop-livestock-forestry system located in Embrapa Beef Cattle (Source: Dr. Ademar P. Serra); and (B) the brand of carbon neutral Brazilian beef (CNBB) concept (Developed by the Brazilian Agricultural Research Corporation (Embrapa Beef Cattle)).

can show a positive budget in mitigation GHG. The use of trees in ICLFS or ILFS has achieved positive budged in sequestrated CO_2 equivalent in integrated system (**Figure 5A**). Brazilian Agricultural Research Corporation (Embrapa beef cattle) launched the concept of neutral carbon meat (NCM) (**Figure 5B**) [50], which are possible to affirm that the trees in ICLFS and ILFS have the capacity to neutralize the entire emission of GHG into this production system.

6. Concluding remarks

The 4R's nutrient stewardship is universal, requiring adjustment in site-specific to improve the N-use efficiency. Based on the 4R's, it is possible to direct the best management practices (BMPs) to achieve sustainable agricultural. The 4R's must be well defined in order to obtain higher N-fertilizer use, consequently decreasing in social, economic, and environmental negative impact, resulting in increased nutrient use efficiency associated with high yielding.

A mixture of *Brachiaria* spp. with legumes must be incentivized to decrease the use of N-sources in forage grasses; however, the competition between species requires more understanding in order to avoid legume degradation in mixture with *Brachiaria* spp.

In order to improve NUE in forage grass, the N-fertilizer must be applied in the right rate, with the right source, followed by the right time and place; this is the sequence that one needs to keep in mind to adjust the 4R's in site-specific to achieve social benefits with the absence of negative environmental impact and improvement of economic returns.

The 4R's are interrelated and N-fertilizer rates cannot be recommended without the combination of the 4R's where a whole system to be followed should be considered to decide about N-fertilizer in pasture. One must keep in mind that the modification of one principal is going to affect the result of the other.

The possibility to sequestrated greenhouse gas in integrated livestock-forestry system (ILFS) is quite important for the environment. As was reported in this chapter, the implementation of trees in integrated systems can neutralize the whole emission of GHG by cattle.

Author details

Ademar Pereira Serra[1]*, Marlene Estevão Marchetti[2], Elisângela Dupas[2], Simone Candido Ensinas[3], Elaine Reis Pinheiro Lourente[2], Eulene Francisco da Silva[4], Roberto Giolo de Almeida[1], Carla Eloize Carducci[2] and Alessandra Mayumi Tokura Alovisi[2]

*Address all correspondence to: ademar.serra@embrapa.br

1 Brazilian Agricultural Research Corporation (EMBRAPA), Campo Grande, Brazil

2 Federal University of Grande Dourados (UFGD), Dourados, Brazil

3 State University of Mato Grosso do Sul (UEMS), Cassilândia, Brazil

4 Federal Rural University of the Semi-arid Region (UFERSA), Mossoró, Brazil

References

[1] McManus C, Barcellos JOJ, Formenton BK, Hermuche PM, Carvalho OA Jr, Guimarães R. Dynamics of cattle production in Brazil. PloS One. 2016;**11**:1-15. DOI: 10.1371/journal.pone.0147138

[2] Euclides VPB, Valle CB, Macedo MCM, Almeida RG, Montagner DB, Barbosa RA. Brazilian scientific progress in pasture research during the first decade of XXI century. Revista Brasileira de Zootecnia. 2010;**39**:151-168

[3] Alexandratos N, Bruinsma J. World Agriculture Towards 2030/2050: The 2012 Revision. Food and Agriculture Organization of the United Nations, ESA Working Paper No. 12-03; June 2012

[4] Monteiro FA. Uso de corretivos e fertilizantes. In: Reis RA, GRS B, Siqueira GR, editors. Forragicultura: Ciência, Tecnologia e Gestão de Recursos Forrageiros. Jaboticabal: Maria de Lourdes Brandel – ME; 2013. p. 275-290

[5] Bruulsema T, Lemunyon J, Herz B. Know your fertilizer rights. Crops and Soils. 2009;**42**:13-18

[6] Martha JGB, Vilela L, Sousa DMG, Barcellos AO. Adubação nitrogenada. In: JGB M, Vilela L, DMG S, editors. Cerrado: uso eficiente de corretivos e fertilizantes em pastagens. 1st ed. Embrapa Cerrados: Planaltina; 2007. p. 117-144

[7] Luo J, Klein CAM, Ledgard SF, Saggar S. Management options to reduce nitrous oxide emissions from intensively grazed pastures: A review. Agriculture, Ecosystems & Environment. 2010;**136**:282-291

[8] Whitehead DC. Nutrient Elements in Grassland: Soil-Plant-Animal Relationship. Oxford: Cabi Publishing; 2000. 363 p

[9] Ensinas SC, Serra AP, Marchetti ME, Silva EF, Prado EAF, Lourente ERP, Altomar PH, Potrich DC, Martinez MA, Conrad VA, Jesus MV, Kadri TCE. Cover crops affects on soil organic matter fractions under no till system. Australian Journal of Crop Science. 2016;**10**:503-512

[10] Ghosh PK, Palsaniya DR, Rai AK, Kumar S. Strategies for higher nutrient use efficiency and productivity in forage crops. In: Rakshit A, Singh HB, Sen A, editors. Nutrient Use Efficiency: From Basics to Advances. 1st ed. India: Springer; 2014. p. 329-342. DOI: 10.1007/978-81-322-2169-2_21

[11] Di HJ, Cameron KC. Reducing environmental impacts of agriculture by using a fine particle suspension nitrification inhibitor to decrease nitrate leaching from grazed pastures. Agriculture, Ecosystems and Environment. 2005;**109**:202-212

[12] Santos FAS, Dórea JRR, Gouvêa VN. Compostos nitrogenados em plantas forrageiras. In: Reis RA, Bernardes GRS, Siqueira GR, editors. Forragicultura: Ciência, Tecnologia e Gestão de Recursos Forrageiros. 1st ed. Jaboticabal: Maria de Lourdes Brandel – ME; 2013. p. 395-407

[13] Schmidt F, Bona FD, Monteiro FA. Sulfur limitation increases nitrate and amino acid pools in tropical forages. Crop and pasture. Science. 2013;**64**:51-60. DOI: 10.1071/CP12336

[14] De Bona FD, Fedoseyenko D, Von Wirén N, Monteiro FA. Nitrogen utilization by sulfur-deficient barley plants depends on the nitrogen form. Environmental and Experimental Botany. 2011;**74**:237-244. DOI: 10.1016/j.envexpbot.2011.06.005

[15] Rowlings DW, Scheer C, Liu S, Grace PR. Annual nitrogen dynamics and urea fertilizer recoveries from a dairy pasture using 15N; effect of nitrification inhibitor DMPP and reduced application rates. Agriculture, Ecosystems & Environment. 2016;**216**:216-225. DOI: 10.1016/j.agee.2015.09.025

[16] Marschner P, Rengel Z. Nutrient Cycling in Terrestrial Ecosystems. Berlin, Germany: Springer-Verlag Berlin; 2007. 291 p

[17] Santos JHS. Proporções de nitrato e amônio na nutrição e produção dos capins Aruana e Marandu [thesis]. Piracicaba: Escola Superior de Agricultura "Luiz de Queiroz", Universidade de São Paulo; 2003

[18] De Bona FD. Nitrogênio e enxofre para gramínea forrageira: Atributos do solo e aspectos metabólicos, nutricionais e produtivos da planta [thesis]. Piracicaba: Escola Superior de Agricultura "Luiz de Queiroz", Universidade de São Paulo; 2008

[19] Marschner P. Marschner's Mineral Nutrition of Higher Plants. 3rd ed. London: Elsevier; 2012. 643 p

[20] Sidhu PK, Bedi GK, Mahajan MV, Sharma S, Sandhu KS, Gupta MP. Evaluation of factors contributing to excessive nitrate accumulation in fodder crops leading to ill-health in dairy animals. Toxicology International. 2011 Jan–Jun;**18**(1):22-26. DOI: 10.4103/0971-6580.75848

[21] Primavesi AC, Primavesi O, Correa LA, Cantarella H, Silva AG, Freitas AR, Vivaldi LJ. Adubação nitrogenada em capim-coastcross: Efeitos na extração de nutrientes e recuperação aparente do nitrogênio. Revista da Sociedade Brasileira de Zootecnia. 2004;**33**:68-78

[22] Dupas E, Buzetti S, Rabelo FHS, Sarto AL, Cheng NC, Teixeira Filho MCM, Galindo FS, Dinalli RP, Gazola RN. Nitrogen recovery, use efficiency, dry matter yield, and chemical composition of palisade grass fertilized with nitrogen sources in the Cerrado biome. Australian Journal of Crop Science. 2016;**10**:1330-1338

[23] Faccin FC, Marchetti ME, Serra AP, Ensinas SC. Frações granulométricas da matéria orgânica do solo em consórcio de milho safrinha com capim-marandu sob fontes de nitrogênio. Pesquisa Agropecuária Brasileira. 2016;**51**:2000-2009

[24] De Bona FD, Monteiro FA. Marandu palisadegrass growth under nitrogen and sulphur for replacing signal grass in degraded tropical pasture. Scientia Agricola. 2010;**67**:570-578. DOI: 10.1590/S0103-90162010000500011

[25] Silveira CP, Nachtigall GR, Monteiro FA. Norms for the diagnosis and recommendation integrated system for signal grass. Scientia Agricola. 2005;**62**:513-519

[26] Andrade CMS, Garcia R, Couto L, Pereira OG, Souza AL. Desempenho de Seis Gramíneas Solteiras ou Consorciadas com o Stylosanthes guianensis cv. Mineirão e Eucalipto em Sistema Silvipastoril. Revista Brasileira de Zootecnia. 2003;**32**:1845-1850

[27] Olanite JA, Tarawali SA, Aken'ova ME. Biomass yield, quality and acceptability of selected grass-legume mixtures in the moist savannah of west Africa. Tropical Grasslands. 2004;**38**:117-128

[28] Reis Junior FB, Silva MF, Teixeira KRS, Urquiaga S, Reis VM. Identification of *Azospirillum amazonense* isolates associated to brachiaria spp. at different stages and growth conditions, and bacterial plant hormone production. Revista Brasileira de Ciência do Solo. 2004;**28**:103-113

[29] Lu X, Fan J, Yan Y, Wang X. Comparison of soil microbial biomass and enzyme activities among three alpine grassland types in northern Tibet. Polish Journal of Environmental Studies. 2013;**22**:437-443

[30] Moore JM, Klose S, Tabatabai MA. Soil microbial biomass carbon and nitrogen as affected by cropping systems. Biology and Fertility of Soils. 2000;**31**:200-2010

[31] Jackson LE, Bowles TM, Hodson AK, Lazcano C. Soil microbial-root and microbial-rhizosphere processes to increase nitrogen availability and retention in agroecosystems. Current Opinion in Environmental Sustainability. 2012;**4**:517-522

[32] Asgedom H, Kebreab E. Beneficial management practices and mitigation of greenhouse gas emissions in the agriculture of the Canadian prairie: A review. Agronomy for Sustainable Development. 2011;**31**:433-451. DOI: 10.1007/s13593-011-0016-2

[33] Bouwman AF, Boumans LJM, Batjes NH. Estimation of global NH3 volatilization loss from synthetic fertilizers and animal manure applied to arable lands and grasslands. Global of Biogeochemical Cycles. 2002;**6**:8-14. DOI: 10.1029/2000GB001389

[34] Soares JR, Cantarella H, Menegale MLC. Ammonia volatilization losses from surface-applied urea with urease and nitrification inhibitors. Soil Biology and Biochemistry. 2012;**52**:82-89

[35] Singh J, Kunhikrishnan A, Bolan NS, Saggar S. Impact of urease inhibitor on ammonia and nitrous oxide emissions from temperate pasture soil cores receiving urea fertilizer and cattle urine. Science of the Total Environment. 2013;**465**:56-63

[36] Di HJ, Cameron KC. Nitrate leaching in temperate agroecosystems: Sources, factors and mitigating strategies. Nutrient Cycling in Agroecosystems. 2002;**46**:237-256

[37] Dupas E, Buzett S, Sarto AL, Hernandez FBT, Bergamaschine AF. Dry matter yield and nutritional value of Marandu grass under nitrogen fertilization and irrigation in Cerrado in Sao Paulo. Revista Brasileira de Zootecnia. 2010;**39**:2598-2603

[38] Snnifen CJ, O'Connor JD, Van Soest PJ, Fox DG, Russel JB. A net carbohydrate and protein system for evaluating cattle diets: II. Carbohydrate and protein availability. Journal of Animal Science. 1992;**70**:3562-3577

[39] Dupas E. Nitrogen potassium and boron: Productive, morphological and nutritional parameters and fibers and protein fractions of Tanzania guinegrass [thesis]. Piracicaba: Escola Superior de Agricultura "Luiz de Queiroz", Universidade de São Paulo; 2012

[40] Oliveira RL, Barbosa MAAF, Cecato U, Matos RC, Rodrigues A, Santiago MS. Carbohydrate and protein fractions of Mombaça grass under different cutting intervals and nitrogen levels. Biotechnology in Animal Husbandry. 2004;**20**:163-167

[41] Johnson CR, Reiling BA, Mislevy P, Hall MB. Effects of nitrogen fertilization and harvest date on yield, digestibility, fiber, and protein fractions of tropical grasses. Journal of Animal Science. 2001;**79**:2439-2448

[42] Inselsbacher E, Wanek W, Ripka K, Hack E, Sessitsch A, Strauss J, Zechmeister-Boltenstern S. Greenhouse gas fluxes respond to different N fertilizer types due to altered plant-soil-microbe interactions. Plant and Soil. 2011;**343**:17-35. DOI: 10.1007/s11104-010-0597-6

[43] IPCC. Climate change 2007: Synthesis report. In: Team CW, Pachauri RK, Reisinger A, editors. Contribution of Working Groups I, II and III to the Fourth Assessment Report of the Intergovernmental Panel on Climate Change. Geneva: IPCC; 2007

[44] Reay DS, Davidson EA, Smith KA, Smith P, Melillo JM, Dentener F, Crutzen PJ. Global agriculture and nitrous oxide emissions. Nature of Climate Change. 2012;**2**:410-416. DOI: 10.1038/nclimate1458

[45] IPCC. Climate change 2007: The physical science basis. In: Solomon S, Qin D, Manning M, Chen Z, Marquis M, Averyt KB, Tignor M, Miller HL, editors. Intergovernmental Panel on Climate Change. Cambridge/New York, NY, UK/USA: Cambridge University Press; 2007. 996 p

[46] Halving JL, Beaton JD, Tisdale SL, Nelson WR. Soil Fertility and Fertilizers: An Introduction to Nutrient Management. 6th ed. NJ: Prentice Hall; 1999 528 p

[47] Zaman M, Saggar S, Blennerhassett J, Singh J. Effect of urease and nitrification inhibitors on N transformation, gaseous emissions of ammonia and nitrous oxide, pasture yield and N uptake in grazed pasture system. Soil Biology and Biochemistry. 2009;**41**:1270-1280

[48] Lessa ACR, Madari BE, Paredes DS, Boddey RM, Urquiaga S, Jantalia CP, Alves BJR. Bovine urine and dung deposited on Brazilian savannah pastures contribute differently to direct and indirect soil nitrous oxide emissions. Agriculture, Ecosystems and Environment. 2014;**190**:104-111

[49] Xu X, Boeckx P, Cleemput OV, Zhou L. Urease and nitrification inhibitors to reduce emissions of CH_4 and N_2O in rice production. Nutrient Cycling in Agroecosystems. 2002;**64**:203-211. DOI: 10.1023/A:1021188415246

[50] Almeida RG, Gomes RC, Silva VP, Alves FV, Feijó GLD, Ferreira AD, Oliveira EB, Bungenstab DJ. Carbon Neutral Brazilian Beef: Testing Its Guidelines through a Case Study. Campo Grande: Embrapa Beef Cattle Doc No. 2016; 2016. 16 p

[51] Baligar VC, Fageria NK, He ZL. Nutrient use efficiency in plants. Communication in Soil Science and Plant Analysis. 2001;**32**:921-950

[52] Fageria NK, Morais OP, Santos AB. Nitrogen use efficiency in upland rice genotypes. Journal of Plant Nutrition. 2010;**33**:1696-1711. DOI: 10.1080/01904167.2010.496892

[53] Gerloff GC, Gabelman WH. Genetic basis of inorganic plant nutrition. In: Lauchli A, Bieleski RL, editors. Inorganic Plant Nutrition. Encyclopedia and Plant Physiology New Series. 15th ed. New York: Springer Verlag; 1983. pp. 453-480

[54] Raun WR, Johnson GV. Improving nitrogen use efficiency for cereal production. Agronomy Journal. 1999;**91**:357-363

Chapter 8

Alfalfa and Its Symbiosis Responses to Osmotic Stress

Mohammed Mouradi, Mohamed Farissi,
Abdelaziz Bouizgaren, Yahya Lahrizi,
Ahmed Qaddoury and Cherki Ghoulam

Additional information is available at the end of the chapter

http://dx.doi.org/10.5772/intechopen.69760

Abstract

Alfalfa (*Medicago sativa* L.) is one of the most cultivated forage legumes in Morocco thanks to its great adaptation to the climatic conditions of this country, its high protein content and its ability to fix atmospheric nitrogen in symbiosis with rhizobia. Environmental stresses such as drought and salinity constitute a major factor limiting the symbiotic nitrogen fixation and legume productivity. In the last decades, this process has interested scholars in understanding the implication of these strains in legume stress tolerance in order to make these symbioses more efficient under difficult conditions. Seed osmopriming is a great technique in the amelioration of seed germination and seedlings growth in responses to several abiotic stress conditions. In this chapter, the effects of water deficit on the Moroccan alfalfa populations and their symbiotic association with rhizobia were discussed. Besides, osmopriming could make these symbioses more efficient especially under stress conditions.

Keywords: alfalfa, drought, salinity, N_2-fixing, osmopriming, photosynthesis

1. Introduction

Alfalfa (*Medicago sativa* L.) is a frequently cultivated legume forage in the Mediterranean region thanks to its high leaf protein content, effects on soil fertility and deep root system [1]. In Morocco, this crop occupies 455,000 ha, which represents about 25% of the area devoted to forage crops where 40% is in irrigated systems. In the southeast oases of Morocco, alfalfa cultivation contributes to socioeconomic life as the main forage crop. Local alfalfa populations have many characteristics of agronomic interest such as tolerance to grazing (capacity for rooting and regrowth and diseases) [2], and are considered moderately tolerant to abiotic stresses, including drought and salinity in these areas, but they show significant variation within many of them depending on their habitats of origin.

Water deficit in these regions is one of the major factors influencing the productivity and persistence of many crops. This stress aggravates the impact of other abiotic or biotic stresses to which plants are exposed. In addition, the increase in water demand by other sectors of the national economy (industry and tourism) and the high incidence of drought due to climate change have led to low water availability for agriculture. In addition, climate change is expected to increase the extent of drought and temporal variation in our Mediterranean region [3]. Moreover, this constraint limits the forage production of alfalfa in several regions of the world via its negative effects on germination, the rate of photosynthesis, the metabolism of the plant and its ability to establish and function in the N_2 fixing symbiosis. A severe and prolonged water deficit decreases the number of rhizobia and affects its genetic diversity in the soil as well as in the rhizosphere [4], by inhibiting the process of root infection and directly influencing the functioning of the nodules and the survival of rhizobia in the soil. As a result, large quantities of chemical fertilizers have to be brought to the soil, some of which (about 1%) are used by plants, while the rest are rapidly converted into insoluble complexes. These lead to the need for frequent application of fertilizers; however, its regular use has become costly and ecologically undesirable, and it raises soil salinity in the long term, hence the need to develop economic and environmentally friendly technologies. The selection, the characterization of drought-tolerant alfalfa populations, their symbiosis and the understanding their responses to theses abiotic stresses are of great importance by taking advantage of the genetic biodiversity of both local populations and rhizobia strains in the soil. Seed osmopriming could be also an effective technique for improving germination and vigour of young seedlings of many species. It is a useful tool to overcome the problems of drought and salinity, ensuring the rapid and successful establishment of seeded seeds and the induction of tolerance mechanisms in young seedlings in post-germination, especially under conditions of stress [5].

In this context, the major objective of this review is dedicated to the presentation of the recent knowledge on the effects of water stress and salinity on the growth and development of alfalfa (*M. sativa* L.), first in the germination stage, during their stage of development and in association with rhizobia isolated from different Moroccan soils. The study was focused on physiological, hydric, growth, biochemical and photosynthesis parameters related to water deficit and salinity tolerance, as well as the presentation of recent knowledge about the impact of seed priming on the tolerance of alfalfa to drought conditions.

2. Alfalfa origins and domestications

It is difficult to recognize the origins of the first domestications of alfalfa (*M. sativa* L.). It was cultivated according to different authors 9000 years ago, in its centre of origin. In Ref. [6], the centre of origin of *M. sativa* is the Near East, Asia Minor, Transcaucasia, Iran and the high areas of Turkmenistan. The most common geographic centre is Iran (**Figure 1**). These regions are characterized by cold winters and dry, warm summers and well-drained and neutral pH soils [7]. In Ref. [8], a second centre of origin, Central Asia, characterized by a dry climate and mild winters was added.

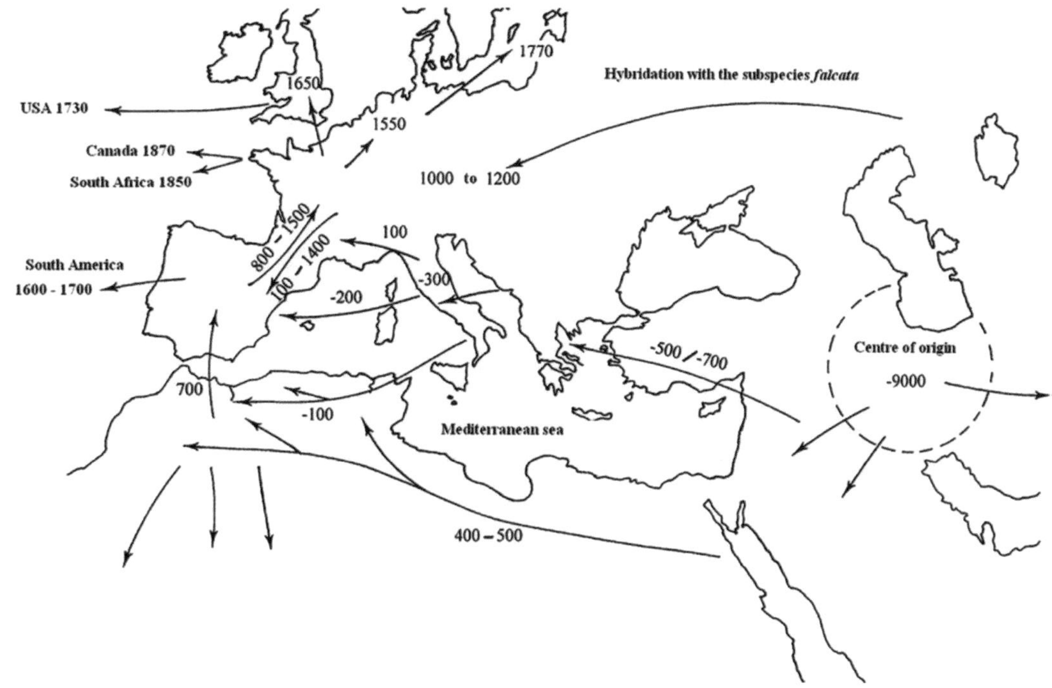

Figure 1. The different routes and approximate dates of the spread of alfalfa cultivated from its centre of origin. (-) Correspond to the dates before JC [11].

Alfalfa cultivation was introduced to North Africa 400–500 years BC through Egypt and 100 years BC by the Roman Empire [9]. These introductions concerned the Atlas Mountains. Other more recent introductions were made in the eighth century by Arab Muslims and concerned the pre-Saharan oases [9], and thanks to Arab Muslims during the seventh and eighth centuries, it was reintroduced into Europe via Africa, and the name "alfalfa" in English is a word from Arabic meaning "al alfa" [10] .العلفة

3. The responses of alfalfa to osmotic stress

3.1. Alfalfa responses of plants to water deficit

The response of alfalfa to water deficit mainly depends on the severity of the stress and growth stage and its physiological state. It results in a 49% decrease in biomass and an 18% increase in leaf-to-stem ratio [12]. The three main mechanisms that reduce the yield of alfalfa due to water deficiency are (i) reduction of the absorption of photosynthetic radiation by the canopy, (ii) decrease of radiation efficiency and (iii) reduction of the harvest index [13].

3.2. Effect on seed germination

Seed germination and seedling growth are the most vulnerable stages to drought. Thus, water stress is one of the main fatalities to seed germination in alfalfa. Germination is a metabolic process requiring three main factors—water, oxygen and temperature—in addition to light as another factor in some species. At seeding, water stress inadequacy delays the germination process and reduces germination percentage and growth rate. It induces irregular germination and non-synchronized emergence of alfalfa seedlings, resulting in low stand populations and reduced yields [14]. In Refs. [1, 15], the exposure of alfalfa seeds to high concentrations of polyethylene glycol (PEG) significantly decreased their germination rate, radicle length and velocity index, and thus the seed germination is inhibited beyond an osmotic pressure of −0.9 MPa. However, under moderate stress, the root length remains intact and even increases in some cases to resist water shortage, and this is probably due to the essential role of roots in the life and function of the plant [1].

3.3. Effect on growth and water uptake

A very moderate water deficit, which does not cause flagrant symptoms, results in a significant change in the morphology and physiology of the plant in many species [16]. It acts negatively on cell division, enlargement and differentiation due to loss of turgor and induces decreased energy supply and synthesis of impaired enzymes. This stress causes, in alfalfa, as in other legumes, reduction of the leaf area, the number of leaves, the closure of the stomata limiting the assimilation of CO_2, photosynthetic activity and growth [17].

The most severe effects of water deficiency occur at the root level and directly affect the water absorption process. The ability of alfalfa root to absorb water is the result of the intrinsic hydraulic properties of the root [18]. Generally, this constraint causes an increase in root/stem ratio due to reduced aerial growth and low root change. This mechanism allows the plant to explore more soil volume to absorb water from deeper layers that are not affected by less developed roots. However, no relationship has been demonstrated between root/stem ratio and water deficit tolerance [19]. In addition, length or root density may have a direct relationship to this tolerance [20].

3.4. Effect on nutrient uptake

Water deficiency negatively affects the nutritional balance of legumes through its adverse effects on assimilation, transport and nutrient distribution. Depending on severity and duration, it reduces the bacterial mineralization of organic matter and negatively affects the ability of the roots to absorb nutrients despite the availability of these nutrients in the soil through its direct effect on root hydraulic conductance and transpiration [21] (**Figure 2**).

3.5. Effect on photosynthesis

The rate of photosynthesis is negatively affected by water deficit due to stomatal and non-stomatal changes [22]. The reduction of stomatal conductance (g_s) and stomatal density is among the remarkable responses to water deficit in alfalfa plants [23, 24]. As a consequence of the stomata closure, the diffusion of CO_2 from the atmosphere to the carboxylation site is reduced, which influences the activity of RuBisCO as well as other enzymes such as sucrose phosphate synthase (SPS) and nitrate reductase (NRA), whose inhibition is often considered to be the main cause of the decrease of water-deficient photosynthesis [25]. It has been shown that decreasing water

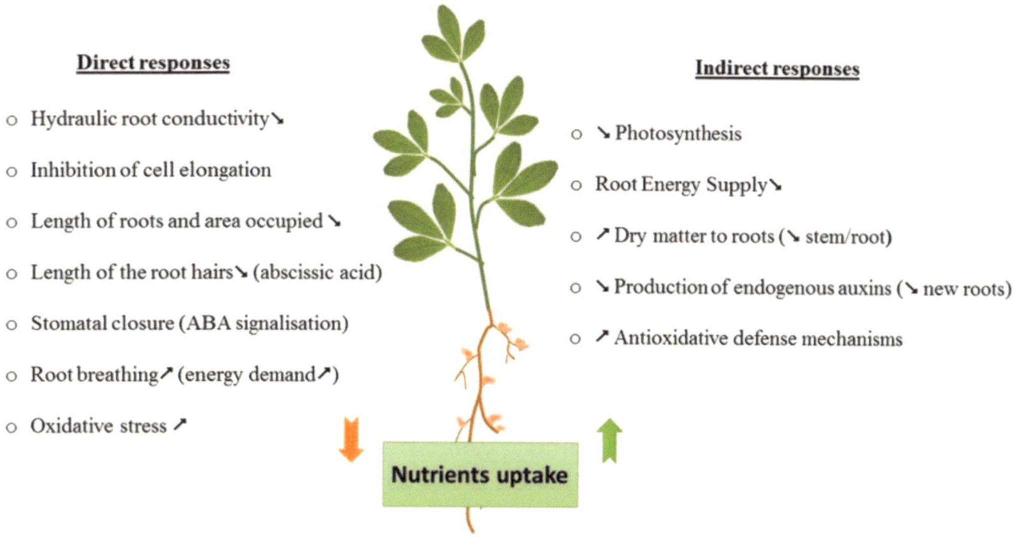

Figure 2. The direct and indirect effects of water deficit on the assimilation of nutrients by roots [21].

content and ion concentration in water-deficient leaves are more limiting for photosynthesis than closure of stomata [26]. Water stress also causes considerable disturbance of photosynthetic pigments in the photosystem (PS) II, leading to degradation of thylakoid membranes [27] and reduction of chlorophyll content, affecting their components [28]. Chlorophyll b (Chl b) is more affected than chlorophyll a (Chl a) [29]. However, fluorescence produced by Chl after excitation by light is a non-destructive and rapid biomarker for the estimation of microbiological and environmental stress responses at the PS II level and its structure and function [30].

4. Alfalfa responses to salinity

Through the enhancement of osmotic pressure, salinity leads to the reduction of water uptake which subsequently affects several other metabolic and physiological processes leading to a prolongation of the germination period [31].

4.1. Effect on growth and nutrient uptake

Salinity affects all physiological processes of the plant. Its effects are mainly reflected by a growth reduction. In Ref. [32], it has shown that salt stress significantly inhibits alfalfa growth. The roots are often more affected than the aerial parts [33].

Salt stress causes an imbalance of the plant mineral nutrition that results from a disruption in the absorption and transport of essential items. In general, the presence of NaCl inhibits of K^+, Ca^{2+}, P_i, NO^{3-} and NH^{4+} uptake and reinforces Na^+ and Cl^- salt ions, which accumulate to become toxic for the plant [32]. Thus, in Ref. [34], salt stress resulted in significant K^+ reductions and Na^+ accumulations in young seedlings of *M. sativa* L.

4.2. Effect on photosynthesis

In response to salt stress, a substantial decrease in a plant's stomatal opening can be observed, but the rates of photosynthesis per unit leaf area sometimes remain unchanged [35]. Following stomatal closure, the internal reduction of CO_2 decreases the activity of several enzymes including RuBisCo [36]. In Ref. [37], it has been found that there is a large reduction in stomatal conductance (g_s) at two genotypes of durum wheat. Thus, limiting carboxylation and reducing the net photosynthetic rate, the effects of salinity on photosynthesis can be caused by alterations in the photosynthetic metabolism or else by secondary effects caused by oxidative stress [36].

4.3. Effect on the cell membrane

The plasma membrane is the main site of interaction between salt and alfalfa plant. Salt stress induces a perturbation of the lipid and protein composition in the cell membrane leading to an uncontrolled electrolyte leakage, thus affecting its permeability and stability [33, 38]. In sugar beet, it has been reported that saline treatment caused a significant leaf electrolyte loss reflecting the disruption and destabilization of cell membranes [38].

4.4. Effect on metabolic activity

Some enzymes with carboxylase activity are influenced by salinity. Indeed, the activity of phosphoenolpyruvate carboxylase (PEPC) and ribulose bisphosphate carboxylase oxygenase (RubisCO), enzymes involved in the fixation of atmospheric carbon dioxide, has been negatively affected by salt stress [39]. This modulation is variable according to the species considered and the stage of development of the plant.

The activities of some enzymes involved in nitrogen nutrition of plants are not immune to turn the effect of salinity. In fact, it turned out that the salt stress has a negative influence on the activities of nitrate reductase and glutamine synthetase [38]. This effect varies according to the species, the variety and its nutrition in nitrogen ions.

5. Alfalfa rhizobia symbiosis under osmotic stress

Alfalfa is one of the legumes capable of fixing large amounts of N_2 (200–400 kg ha^{-1} yr^{-1}) in symbiosis with *Ensifer* (*Sinorhizobium*), and drought can indirectly reduce its productivity through impaired performance of the nodules. This stress affects not only the biomass of the nodules but also their functioning. In addition, several studies have reported that biological N_2 fixing (BNF) is conditioned by the active interaction of nodules with the leaves and roots [40–42].

The effect of salinity on symbiosis manifests at different levels. Indirectly, salt stress can affect this symbiosis by reducing the growth of the host plant and affecting some of its physiological processes or directly by inhibiting the infection process and nodule development [43]. Under saline conditions, many studies have reported that the activity of the key enzyme in the N_2 reduction process, nitrogenase, is greatly reduced [44].

6. Effect on multiplication and survival of rhizobia

Drought is one of the important factors influencing the proliferation of soil microflora. It reduces the availability of water around the soil particles and increases the salt concentration in the soil solution, which subsequently leads to adverse effects on growth, rhizobia persistence, movement and ultimately and their diversity in the soil [45]. Water deficiency has been shown to reduce the survival of almost all species of rhizobia, whether or not capable of nodulating legumes and often influencing the map of genetic diversity of different species in the soil. Rapidly growing rhizobia is the most affected in comparison with slow-growing rhizobia [46]. Variability of the genetic tolerance potential has been observed in several rhizobia species such as *Sinorhizobium* [47]. The survival of these water-deficient strains mainly depends on their ability to enter into symbiosis even if this symbiosis is not very effective [48].

7. Effect on alfalfa plant in symbiosis

The detrimental effect of drought on BNF is manifested at several levels of symbiotic interaction, namely, the early stages of infection during development and the functioning of the nodules. The formation of new root hairs and the lengthening of previously dissociated hairs are reduced in response to water deficiency which results in a strong reduction in the plant-bacterial interaction as well as the formation of the infection cord [49]. The aerial and nodular biomass also shows a considerable reduction under these conditions, followed by a reduction in the efficiency of the nitrogenase complex for BNF [50]. This reduction cannot be mainly explained by the decrease in the rate of photosynthesis, whereas other alternative causes can regulate the BNF under water deficit. In Refs. [51, 52], O_2 limitation, C shortage and N-feedback are the three main factors that could be involved in this inhibition. Indeed, severe water deficiency disrupts the nitrogenase activity of the nodules by causing a rapid decrease in the supply of oxygen to the nodules and consequently an interruption of the adenosine triphosphate (ATP) supply to the nitrogenase [53]. It can also slowdown the transport of photosynthate towards the nodules, which limits the contribution of the energy substrates [54].

In Ref. [55], it has been reported that, in general, rhizobia strains appear to be more tolerant to water deficiency than host plants, but these strains exhibit variation in their growth and survival under this stress. The reduction of the nodular biomass under water deficit could be explained by the reduction in the number and diameter of the root hairs or the inhibition of the emergence and the elongation of these organs [56] and, on the other hand, Limited rhizobia growth, which reduces the initiation and development of nodules [57]. Studies reported that g_s was reduced in all of the studied symbiotic combinations subject to water deficiency. This confirms the results found in Refs. [42, 58]. The fact that these combinations exhibited lower g_s combined with high forage yield even under limiting water conditions supports the hypothesis that tolerant cultivars should develop mechanisms for better use of fixed CO_2 and that these mechanisms are related to day/night differential control [59].

In Ref. [24], it has been reported that no significant correlation was observed between leaf area and relative water content (RWC) and between g_s and RWC, while leaf area was found to be highly correlated ($r^2 = 0.674$) at g_s. Thus, we suggest that this could have positive indirect effects on RWC by controlling stomata behaviour, maintaining the leaf area and facilitating water uptake in the event of water deficit [60]. The rate of electrolyte loss is considered to be a good physiological index which reflects the degree of alteration of the plants in the plasma membranes of the cells under stressful conditions [38]. An increase in electrolyte loss indicates that the integrity of the membrane is affected. The results of our study showed a significant increase of this parameter under water deficit in all the studied symbiotic combinations. However, this increase was not significant for the more tolerant combination *Ad-RcRh09* which proves the importance of this parameter in the osmotic stress tolerance [24]. The same results have been reported in Refs. [61, 62].

8. Drought tolerance mechanisms in alfalfa

8.1. Osmotic adjustment and water potential

Water potential (Ψ) and osmotic adjustment (AO) can be used as selection criteria to improve drought tolerance in many legume species grown in arid and semiarid regions. Maintaining a low Ψ is considered one of dehydration avoidance mechanisms developed by plants in order to survive in extreme drought conditions. It depends on access and absorption of soil water by roots, sweating, canopy size, leaf area, leaf rolling and internal water transport [63]. On the other hand, substances that can contribute to AO include organic acids, cations (such as K^+) and inorganic anions, carbohydrates and amino acids. These substances are often associated with easy protective functions such as hydroxyl (OH), cyclitol, proline and glycine betaine [64]. Active AO is the net increase in osmotically active solutes leading to the decrease of the osmotic potential (Ψ_s) in the cell and consequently the total Ψ decrease. This is seen as a drought adaptation and not a mere response. On the other hand, a second form of AO has been proposed according to the Ref. [65], called "passive AO", considered as a response to water deficit and associated with loss of water and therefore a reduction of the cell volume.

8.2. Antioxidant defence

Water deficiency induces the appearance of oxidative stress, that is, the accumulation of reactive oxygen species (ROS) causing damage to cellular structures [66]. These are molecules showing redox states between oxygen (O_2) and water (H_2O) including superoxide anion O_2^-, hydrogen peroxide (H_2O_2), hydroxide (OH^-) and oxygen singlet ($^1O_2$) which are extremely reactive and tend to reduce the water molecule rapidly (ms to ps). As a result, legumes have developed a powerful antioxidant system that is finely regulated in time and space to maintain adequate levels of ROS. This system could detoxify the ROS and radicals of lipid peroxides and maintain an adequate redox balance in the cell [40]. An antioxidant is defined as a molecule capable of releasing H^+ electrons or protons with a low reduction potential in order to have a radical either harmless or effectively quenched by other electron donors and have properties that correlate with oxidative stress [67]. Numerous studies have shown the induction of nonenzymatic and enzymatic antioxidants in legumes subjected to water deficiency. Antioxidant enzymes are proteins that catalyse the detoxification of free radicals by using protons or electrons released by nonenzymatic antioxidants [68]. In the case of water deficit and salinity, several workers have reported the increase of these activities in the nodules under stress [41]. It has been shown in Refs. [40, 69] that tolerant alfalfa-rhizobia symbiotic combinations synthesize more peroxidase, catalase and superoxide dismutase in their leaves and nodules than the sensitive ones.

9. Deficit irrigation and water productivity (WP)

Irrigation schemes can be classified as full and deficit irrigation regimes, depending on the crop species, the physiological conditions of the plant, the soil and the climate of the region in question. With full field capacity irrigation, high growth and stable yield could be achieved but require high amounts of water and high cost of accompanying farming practices. The water requirements of each crop vary at different stages of plant development and genotype, sensitivity of its physiological state, soil structure and property, climatic conditions and agricultural practices [28]. To avoid water stress damage during the growth phase, deficit irrigation could be an additional method to improve water productivity (WP) in alfalfa and other crops especially in areas where water resources are limited or production costs are high [70], as in the case of the arid regions of Morocco where the cultivation of lucerne is conducted only by irrigation. It has been reported in Refs. [71, 72] that deficit irrigation could improve the WP and hence the growth and yield of plants relative to full irrigation. However, in Ref. [73] summer deficit irrigation reduces the yield of alfalfa without impeding its growth.

10. Seed osmopriming and drought tolerance in alfalfa

Seed priming with chemical agents such as sodium nitroxide, hydrogen peroxide, sodium hydrosulphide, melatonin, polyamines and polyethylene glycol (PEG) or biological agents such as bacterial suspensions improves plants tolerance to different abiotic stresses by improving cellular homeostasis and plant growth [74]. The most commonly used priming agents share the same modes of action, especially under stressful conditions. Moreover, when used against different abiotic constraints, their modes of action exhibit similarities, but also distinct specificities and their performance mainly depend on concentration of the priming agent, priming period and temperature [74].

The purpose of seed osmopriming is to reduce germination time and to improve germination percentage especially under adverse environmental conditions. Treated seeds have been shown to have the potential to rapidly restart germinative metabolism, thereby improving germination rates [75]. The impact of abiotic stress on the physiology and growth of alfalfa, maize and soybean plants from treated seeds has been shown to be remarkably reduced compared to plants from untreated seeds [15, 33] (**Figure 3**). Moreover, those plants whose seeds

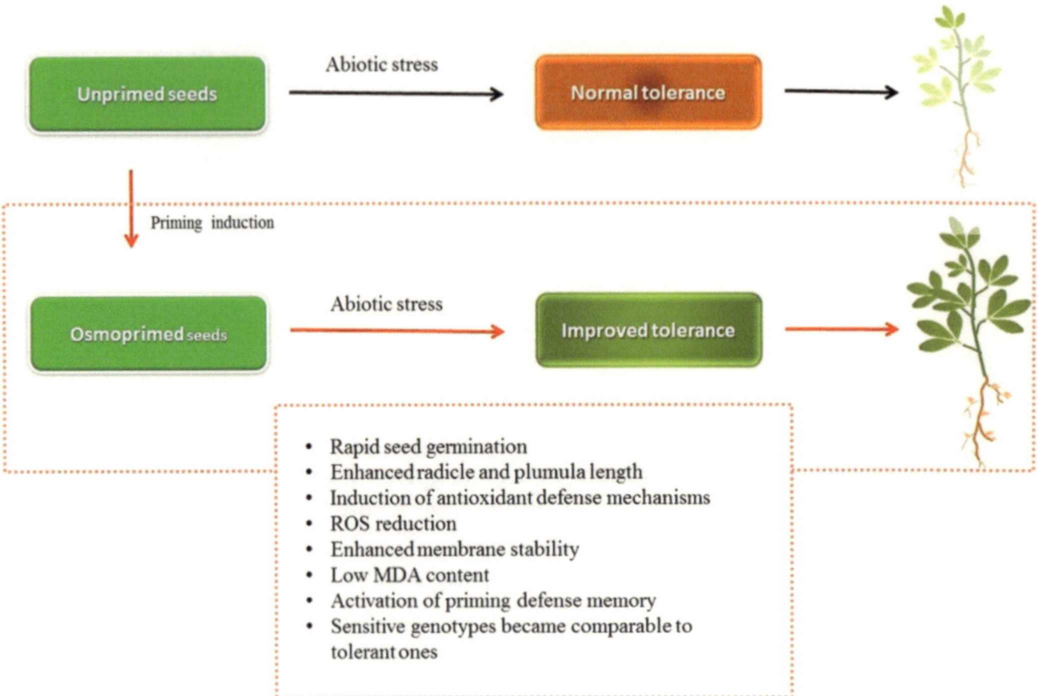

Figure 3. Seed osmopriming as a way to improve drought tolerance in alfalfa.

have been previously exposed to a pretreatment agent such as a natural or synthetic chemical compound present opportunities for better use in the study and management of the physiology of biotic and abiotic stresses in plants.

10.1. Osmopriming and pregerminative metabolism

In the osmotic priming technique, seeds are soaked in PEG or other osmotic solution during the period of time estimated to complete the first two phases of the germination process. Several osmotic agents can be used, mainly KNO_3, KH_2PO_4, K_2HPO_4, $CaCl_2$, $ZnSO_4$, $MgCl_2$, $MnSO_4$, NaCl, $NaSO_4$ and organic compounds, namely, fumaric, succinic, malic and citric acids; purines; and pyrimidines. PEG is most commonly used as a Ψ reducing agent due to its non-toxic nature and large molecular size, without penetrating the seeds during soaking [76]. This technique has beneficial effects on plant germination and tolerance, especially under osmotic stress conditions. During the treatment, the amount of absorbed water is controlled in such a way to induce the pregerminative metabolic activities necessary for germination but prevents the actual emergence of the radicle [77]. Different physiological and biochemical activities occur in the seed at different moisture levels. Generally, this treatment improves seed rate, uniformity and germination time [78]. This effect can be attributed to the activation of seed repair mechanisms after exposure to adverse conditions, accumulation of germination-promoting substances, nutrient accumulation and osmotic adjustment (AO) [78, 79]. This technique has been shown to be strongly associated with increased antioxidant defence in germinated seeds, which allows better tolerance to oxidative stress, reduction of lipid peroxidation and increase in membrane stability under water deficit conditions [15, 80].

10.2. Other techniques of seed priming

There are several techniques for seed priming, which differ according to the used agent. Among the most frequently used are hydropriming and biopriming. The first technique consists in exposing the seeds to a limited quantity of water in a continuous or successive manner at a suitable temperature. This is an inexpensive technique often used for field-grown cereals and legumes to accelerate germination [81]. The second technique consists in treating the seeds with microbiological agents such as rhizobia, *Azospirillum*, *Pseudomonas*, *Bacillus*, *Trichoderma*, *Gliocladium* and other species and aims to improve the vigour and viability of the seeds. This technique uses a combination of hydration and seed inoculation with beneficial organisms in order to improve their germination, especially to protect them under stressful environmental conditions [77].

10.3. Osmopriming drought tolerance

The beneficial effects of seed priming to improve the germination rate under abiotic stress have been reported in some alfalfa genotypes [15, 33] and other legumes such as faba bean [82] and soybeans [83] and other crops such as cumin and rice [84]. At the molecular level, seed priming may be strongly linked to tolerance to water deficit. Indeed, it has been proposed in Refs. [85] that priming involves the accumulation of inactive cell kinase cascades and the modification of the chromatin structure and thus allows the amplification and activation of stress defence genes. In Ref. [86], two strategies have been proposed in which osmopriming probably improves abiotic stress tolerance including water deficit and salinity. In the first place, the treated seeds mobilize activities related to germination, for example, respiration, weakening of the endosperm, transcription and translation of genes, etc., thus facilitating the transition of dry seeds from the state of resting towards germination and leading to the improvement of their germination potential. Secondly, osmopriming initially imposes a certain level of abiotic stress on the seeds that suppresses the emergence of the radicle but stimulates stress-related reactions such as the accumulation of abundant proteins of late embryogenesis (LEA). These two strategies constitute a sort of priming memory, which could participate in the mechanisms of tolerance of germinated seeds during subsequent stress exposure [87]. In Ref. [15], priming of alfalfa seeds at −0.6 MPa of PEG6000 for 24 h at 25°C improved the water deficit tolerance in germinated seeds. Similar results have been reported in Refs. [84, 88] for rice and cumin, respectively.

In general, tolerant genotypes have the ability to protect themselves by stimulating the synthesis of enzymes and antioxidant molecules [89]. These compounds can thus neutralize the toxic capacity of the peroxide, superoxide and the hydroxyl radicals present in the tissues [90]. In our studies, we observed a significant increase in peroxidase (PO) and catalase (CAT) activities under water deficiency in young seedlings of all tested alfalfa genotypes [15]. These increases showed relative variations between genotypes. In addition, CAT activity was positively correlated ($r = 0.161^*$) to germination performance under water deficiency, which could be explained by the induction of CAT enzyme synthesis, which plays a key role in the protection and repair systems under water deficiency, especially during PEG priming [91]. The results showed that priming improves membrane protection in most alfalfa seedlings under severe water stress (−0.75 MPa PEG), especially in seedlings treated with −0.6 MPa PEG6000. Similar results have been reported in Refs. [92] after 48 h of osmopriming for spinach seeds and after 12 h of osmopriming in [33] for alfalfa under saline stress. Under similar conditions, the *Adis-Tata* population showed the greatest stability of the membrane compared to the other genotypes studied. It presented low levels of malondialdehyde (MDA) and low electrolyte loss values in comparison to other tested populations, and these results could be explained by the high levels of antioxidant enzyme activity such as PO and CAT [90]. The low accumulation of MDA in tolerant cultivars could be explained by the decomposition of ROS via increased CAT and PO activities and is consistent with improved protection of some *Medicago* cultivars against oxidative damage [41].

10.4. Osmopriming and N_2-fixing symbiosis

Seed osmopriming is a good technique to improve the germination rate and vigour of young seedlings as well as the plant growth in many species. It has been shown that it can improve nodulation, N_2 capacity in legumes and nutrient acquisition, especially in less fertile soils [81]. Studies have reported that plants from treated seeds have shown high capacity to form nodules and accumulate large amounts of N, K^+ and P, especially under stress conditions [93, 94]. In addition, rapid germination of seedlings could emerge and produce deep roots before the upper soil layers are dried and crusty, which may result in better legume placement and improve their ability to form a large number of nodules [93]. This technique has been reported to be effective in improving growth and yield of legumes [95]. It has been showed in Ref. [96] that the application of *Rhizobium* with seed priming significantly increased photosynthesis and nodulation and consequently nitrogenase activity in *Cicer arietinum*. It has been suggested that the combination of a tolerant rhizobia with some stress-tolerant alfalfa cultivars could improve the ability of plants to grow and survive under water and salinity deficiency conditions [15, 93].

In Ref. [93], osmopriming increased significantly ($p<0.001$) the chlorophyll-fluorescence (Fv/Fm) ratio, time to maximal fluorescence (TFm) and electron transport rate (ETR) in almost all symbiotic associations subjected to water deficit. These results indicated that this treatment may reduce the adverse effects of water deficit and salinity in alfalfa plants [97]. Several traits in these studies such as high chlorophyll contents, ETR and leaf area (canopy), could be behind the improvement of photosynthesis efficiency in combination with osmoprimed seeds in comparison to those from unprimed ones. In addition, we suggest that osmoprimed seeds and improved symbiotic N_2 fixing, high leaf relative water content (RWC) and photosystem (PS) II efficiency in the tolerant symbiotic combinations could avoid leaf senescence under water stress.

Under water deficiency, it has been reported that nutrient level was reduced in alfalfa plants from unprimed seeds in comparison to those from osmoprimed seeds [93]. However, the K^+ content was significantly ($p < 0.001$) increased in tolerant combinations from the treated seeds compared to those of untreated seeds. Seed priming has been shown to intensify seed supply consumption, depletion rates, and dry seedling biomass [93, 98]. In addition, a high absorption of nutrients depends on good seed germination, vigorous establishment, root growth and activation of tolerance mechanisms such as osmoregulators and ROS detoxification enzymes [15, 40].

11. Conclusion

The negative effects of osmotic stress on legumes could strongly determine the interaction between rhizobia and alfalfa symbioses. These conditions have negative effects on plant metabolism and photosynthesis. However, several symbiotic interactions between alfalfa and tolerant rhizobia have shown high tolerance to drought and salinity with significant variation in their behaviour. Osmopriming treatment improves water deficit tolerance in young seedlings of alfalfa as well as the N_2 fixing capacity in growth stage. This enhancement is strongly related to the induction of antioxidant enzymes and due to also to the presence of tolerant rhizobia. This technique is very effective for the less tolerant genotypes and could make them comparable to the tolerant ones under water deficit.

Author details

Mohammed Mouradi[1,2]*, Mohamed Farissi[3], Abdelaziz Bouizgaren[2], Yahya Lahrizi[1], Ahmed Qaddoury[1] and Cherki Ghoulam[1]

*Address all correspondence to: mohammed.mouradi@edu.uca.ma

1 Unit of Biotechnology and Symbiosis Agrophysiology, Faculty of Sciences and Techniques, Cadi Ayyad University, Marrakesh, Morocco

2 Unit of Plant Breeding, National Institute for Agronomic Research (INRA), Marrakech, Morocco

3 Polyvalent Laboratory on Research & Development, Polydisciplinary Faculty, Beni Mellal, Morocco

References

[1] Hamidi H, Safarnejad A. Effect of drought stress on alfalfa cultivars (*Medicago sativa* L.) in germination stage. American-Eurasian Journal of Agricultural and Environmental Science. 2010;**8**(6):705-709

[2] Bouizgaren A. Fiche technique sur la culture de la luzerne au Maroc. technique de production fourragère et semencière. [Technical sheet of alfalfa cultivation in Morocco. Techniques of forage and seed production]. Marrakech, Morocco: National Institute for Agronomical Research (INRA); 2007. pp. 5-23

[3] IPCC, Climate Change 2014: Impacts, Adaptation, and Vulnerability. Part A: Global and Sectoral Aspects. Contribution of Working Group II to the Fifth Assessment Report of the Intergovernmental Panel on Climate Change. Cambridge, United Kingdom and New York, NY, USA: Cambridge University Press; 2014

[4] Hameed RA, Hussain NN. Genetic diversity of rhizobia nodulating alfalfa in Iraq as a source of more efficient drought tolerance strains. Journal of Molecular Biology Research. 2016;**6**:20

[5] Yacoubi RC, Job M, Belghazi W, Chaibi D. Job, Proteomic analysis of the enhancement of seed vigour in osmoprimed alfalfa seeds germinated under salinity stress. Seed Science Research. 2013;**23**:99-110

[6] Vavilov NI, Freier F. Studies on the origin of cultivated plants. Studies on the Origin of Cultivated Plants. 1951. pp.xxii + 185 pp

[7] Michaud R, Lehman W, Rumbaugh M. World distribution and historical development. Alfalfa and Alfalfa Improvement. 1988:25-91

[8] Sinskaya E. Data on the developmental biology and physiology of medic species. Flora of Cultivated Plants of the USSR. 1961:209-291

[9] Birouk A. Les ressources phytogénétiques au Maroc: analyse de la variabilité génétique d'une espèce fourragère, la luzerne (*Medicago sativa* L.). Thèse de doctorat Sciences biologiques Paris, 11;1987:1987PA112377. http://www.theses.fr/1987PA112377

[10] Bouizgaren A. Evaluation de la variabilité génétique à l'aide des caractères d'adaptation et moléculaires chez les populations marocaines de luzerne (*Medicago sativa* L.): Apport à la conservation in-situ. Docteur: Cadi Ayyad University; 2012

[11] Prosperi JM, Jenczewski E, Muller MH, Fourtier S, Sampoux JP, Ronfort J. Alfalfa domestication history, genetic diversity and genetic resources. Legume Perspectives. 2014;**4**:13-14

[12] Bouizgaren A, Farissi M, Ghoulam C, Kallida R, Faghire M, Barakate M, Al Feddy MN. Assessment of summer drought tolerance variability in Mediterranean alfalfa (*Medicago sativa* L.) cultivars under Moroccan fields conditions. Archives of Agronomy and Soil Science. 2013;**59**:147-160

[13] Earl HJ, Davis RF. Effect of drought stress on leaf and whole canopy radiation use efficiency and yield of maize. Agronomy Journal. 2003;**95**:688-696

[14] Wu C, Wang Q, Xie B, Wang Z, Cui J, Hu T. Effects of drought and salt stress on seed germination of three leguminous species. African Journal of Biotechnology. 2011;**10**:17954-17961

[15] Mouradi M, Bouizgaren A, Farissi M, Makoudi B, Kabbadj A, Very AA, Sentenac H, Qaddoury A, Ghoulam C. Osmopriming improves seeds germination, growth, antioxidant responses and membrane stability during early stage of Moroccan alfalfa populations under water deficit. Chilean Journal of Agricultural Research. 2016;**76**:265-272

[16] Tardieu F, Parent B, Caldeira CF, Welcker C. Genetic and physiological controls of growth under water deficit. Plant Physiology. 2014;**164**:1628-1635

[17] Yousfi N, Sihem N, Ramzi A, Abdelly C. Growth, photosynthesis and water relations as affected by different drought regimes and subsequent recovery in *Medicago laciniata* (L.) populations. Journal of Plant Biology. 2016;**59**:33-43

[18] Knipfer T, Fricke W. Water uptake by seminal and adventitious roots in relation to whole-plant water flow in barley (*Hordeum vulgare* L.). Journal of Experimental Botany. 2011:**62**;717-733

[19] Zollinger N, Kjelgren R, Cerny-Koenig T, Kopp K, Koenig R. Drought responses of six ornamental herbaceous perennials. Scientia Horticulturae. 2006:**109**;267-274

[20] Jongrungklang N, Toomsan B, Vorasoot N, Jogloy S, Boote K, Hoogenboom G, Patanothai A. Rooting traits of peanut genotypes with different yield responses to pre-flowering drought stress. Field Crops Research. 2011;**120**:262-270

[21] Rouphael Y, Cardarelli M, Schwarz D, Franken P, Colla G. Effects of drought on nutrient uptake and assimilation in vegetable crops. In: Plant Responses to Drought Stress. Springer Berlin, Heidelberg, Germany; 2012. pp. 171-195

[22] Flexas J, Gallé A, Galmés J, Ribas-Carbo M, Medrano H. The response of photosynthesis to soil water stress. In: Plant Responses to Drought Stress. Springer Ricardo A (Ed.). Berlin; 2012. pp. 129-144

[23] del Pozo A, Ovalle C, Espinoza S, Barahona V, Gerding M, Humphries A. Water relations and use-efficiency, plant survival and productivity of nine alfalfa (*Medicago sativa* L.) cultivars in dryland Mediterranean conditions. European Journal of Agronomy. 2017;**84**:16-22

[24] Mouradi M, Farissi M, Bouizgaren A, Makoudi B, Kabbadj A, Very AA, Sentenac H, Qaddourya A, Ghoulam C. Effects of water deficit on growth, nodulation and physiological and biochemical processes in Medicago sativa-Rhizobia Symbiotic Association. Arid Land Research and Management. 2016;**30**:193-208

[25] Samarah N, Alqudah A, Amayreh J, McAndrews G. The effect of late-terminal drought stress on yield components of four barley cultivars. Journal of Agronomy and Crop Science. 2009:**195**;427-441

[26] Tang AC, Kawamitsu Y, Kanechi M, Boyer JS. Photosynthetic oxygen evolution at low water potential in leaf discs lacking an epidermis. Annals of Botany. 2002;**89**:861-870

[27] Kannan N, Kul G. Drought induced changes in physiological, biochemical and phytochemical properties of Withania somnifera Dun. Journal of Medicinal Plants Research. 2011;**5**:3929-3935

[28] Mathobo R, Marais D, Steyn JM. The effect of drought stress on yield, leaf gaseous exchange and chlorophyll fluorescence of dry beans (*Phaseolus vulgaris* L.). Agricultural Water Management. 2017;**180**:118-125

[29] Jaleel CA, Manivannan P, Wahid A, Farooq M, Al-Juburi HJ, Somasundaram R, Panneerselvam R. Drought stress in plants: A review on morphological characteristics and pigments composition. International Journal of Agriculture and Biology. 2009;**11**:100-105

[30] Lichtenthaler HK, Ac A, Marek MV, Kalina J, Urban O. Differences in pigment composition, photosynthetic rates and chlorophyll fluorescence images of sun and shade leaves of four tree species. Plant Physiology and Biochemistry. 2007;**45**:577-588

[31] Ramin A. Effects of salinity and temperature on germination and seedling establishment of sweet basil (*Ocimum basilicum* L.). Journal of Herbs, Spices & Medicinal Plants. 2006;**11**:81-90

[32] Mezni M, Albouchi A, Bizid E, Hamza M. Minerals uptake, organic osmotica contents and water balance in alfalfa under salt stress. Journal of Phytology. 2010;**2**

[33] Amooaghaie R. The effect of hydro and osmopriming on alfalfa seed germination and antioxidant defenses under salt stress. African Journal of Biotechnology. 2011;**10**:6269-6275

[34] Farissi M, Bouizgaren A, Faghire M, Bargaz A, Ghoulam C. Agro-physiological responses of Moroccan alfalfa (*Medicago sativa* L.) populations to salt stress during germination and early seedling stages. Seed Sciences & Technology. 2011;**39**:389-401

[35] Munns R, Tester M. Mechanisms of salinity tolerance. Annual Review of Plant Biology. 2008;**59**:651-681

[36] Chaves MM, Flexas J, Pinheiro C. Photosynthesis under drought and salt stress: Regulation mechanisms from whole plant to cell. Annals of Botany. 2009;**103**:551-560

[37] James RA, Rivelli AR, Munns R, Von Caemmerer S. Factors affecting CO_2 assimilation, leaf injury and growth in salt-stressed durum wheat. Functional Plant Biology. 2002;**29**:1393-1403

[38] Ghoulam C, Foursy A, Fares K. Effects of salt stress on growth, inorganic ions and proline accumulation in relation to osmotic adjustment in five sugar beet cultivars. Environmental and Experimental Botany. 2002;**47**:39-50

[39] Faghire M, Bargaz A, Farissi M, Palma F, Mandri B, Lluch C, García NT, Herrera-Cervera J, Oufdou K, Ghoulam C. Effect of salinity on nodulation, nitrogen fixation and growth of common bean (*Phaseolus vulgaris*) inoculated with rhizobial strains isolated from the Haouz region of Morocco. Symbiosis. 2011;**55**:69-75

[40] Mouradi M, Bouizgaren A, Farissi M, Qaddoury A, Ghoulam C. Medicago sativa-rhizobia symbiosis under water deficit: Physiological, antioxidant and nutritional responses in nodules and leaves. Journal of Plant Nutrition. In press. 2017

[41] Sharma P, Jha AB, Dubey RS, Pessarakli M. Reactive oxygen species, oxidative damage, and antioxidative defense mechanism in plants under stressful conditions. Journal of Botany. 2012;**2012**:1-26

[42] Aranjuelo I, Molero G, Erice G, Avice JC, Nogués S. Plant physiology and proteomics reveals the leaf response to drought in alfalfa (*Medicago sativa* L.). Journal of Experimental Botany. 2011;**62**:111-123

[43] Arrese-Igor C, González EM, Marino D, Ladrera R, Larrainzar E, Gil-Quintana E. Physiological responses of legume nodules to drought. Plant Stress. 2011;**5**:24-31

[44] Aydi S, Sassi S, Abdelly C. Growth, nitrogen fixation and ion distribution in Medicago truncatula subjected to salt stress. Plant and Soil. 2008;**312**:59

[45] Djedidi S, Yokoyama T, Tomooka N, Ohkama-Ohtsu N, Risal CP, Abdelly C, Sekimoto H. Phenotypic and genetic characterization of rhizobia associated with alfalfa in the Hokkaido and Ishigaki regions of Japan. Systematic and Applied Microbiology. 2011:**34**:453-461

[46] Mhadhbi H, Fotopoulos V, Djebali N, Polidoros AN, Aouani ME. Behaviours of medicago truncatula-sinorhizobium meliloti symbioses under osmotic stress in relation with the symbiotic partner input: Effects on nodule functioning and protection. Journal of Agronomy and Crop Science. 2009;**195**:225-231

[47] Gehlot HS, Panwar D, Tak N, Tak A, Sankhla IS, Poonar N, Parihar R, Shekhawat NS, Kumar M, Tiwari R. Nodulation of legumes from the Thar desert of India and molecular characterization of their rhizobia. Plant and Soil. 2012;**357**:227-243

[48] Sprent JI. Evolution and diversity in the legume-rhizobium symbiosis: Chaos theory? Plant and Soil. 1994;**161**:1-10

[49] Mhadhbi H, Fotopoulos V, Mylona PV, Jebara M, Elarbi Aouani M, Polidoros AN. Antioxidant gene-enzyme responses in Medicago truncatula genotypes with different degree of sensitivity to salinity. Physiologia Plantarum. 2011;**141**:201-214

[50] Oldroyd GE, Murray JD, Poole PS, Downie JA. The rules of engagement in the legume-rhizobial symbiosis. Annual Review of Genetics. 2011;**45**:119-144

[51] Marino D, Damiani I, Gucciardo S, Mijangos I, Pauly N, Puppo A. Inhibition of nitrogen fixation in symbiotic Medicago truncatula upon Cd exposure is a local process involving leghemoglobin. Journal of Experimental Botany. 2013:ert334

[52] González EM, Larrainzar E, Marino D, Wienkoop S, Gil-Quintana E, Arrese-Igor C. Physiological responses of N_2-fixing legumes to water limitation. In: Legume Nitrogen Fixation in a Changing Environment. Springer, New York; 2015. pp. 5-33

[53] Larrainzar E, Wienkoop S, Scherling C, Kempa S, Ladrera R, Arrese-Igor C, Weckwerth W, González EM. Carbon metabolism and bacteroid functioning are involved in the regulation of nitrogen fixation in Medicago truncatula under drought and recovery, Molecular Plant-Microbe Interactions. 2009;**22**:1565-1576

[54] Serraj R, Vadez V, Sinclair T. Feedback regulation of symbiotic N_2 fixation under drought stress. Agronomie. 2001;**21**:621-626

[55] Ben Romdhane S, Trabelsi M, Aouani ME, de Lajudie P, Mhamdi R. The diversity of rhizobia nodulating chickpea (*Cicer arietinum*) under water deficiency as a source of more efficient inoculants. Soil Biology and Biochemistry. 2009;**41**:2568-2572

[56] Zaharan H, Sprent J. Effect of sodium chloride and polyethylene glycol on root hair infection of *Vicia faba* L. plants by *Rhizobium leguminosarum*. Planta. 1986;**167**:303-309

[57] Antolín MC, Muro I, Sánchez-Díaz M. Application of sewage sludge improves growth, photosynthesis and antioxidant activities of nodulated alfalfa plants under drought conditions. Environmental and Experimental Botany. 2010;**68**:75-82

[58] Leung J, Valon C, Moreau B, Boeglin M, Lefoulon C, Joshi-Saha A, Chérel I. Mécanisme moléculaire d'action de l'acide abscissique en réponse à la sécheresse chez les végétaux. Biologie Aujourd'hui. 2012;**206**:301-312

[59] Rosales MA, Ocampo E, Rodríguez-Valentín R, Olvera-Carrillo Y, Acosta-Gallegos J, Covarrubias AA. Physiological analysis of common bean (*Phaseolus vulgaris* L.) cultivars uncovers characteristics related to terminal drought resistance. Plant Physiology and Biochemistry. 2012;**56**:24-34

[60] Thameur A, Lachiheb B, Ferchichi A. Drought effect on growth, gas exchange and yield, in two strains of local barley Ardhaoui, under water deficit conditions in southern Tunisia. Journal of Environmental Management. 2012;**113**:495-500

[61] Slama I, Tayachi S, Jdey A, Rouached A, Abdelly C. Differential response to water deficit stress in alfalfa (*Medicago sativa*) cultivars: Growth, water relations, osmolyte accumulation and lipid peroxidation, African Journal of Biotechnology. 2011;**10**:16250-16259

[62] Farissi M, Ghoulam C, Bouizgaren A. Changes in water deficit saturation and photosynthetic pigments of Alfalfa populations under salinity and assessment of proline role in salt tolerance. Agricultural Science Research Journals. 2013;**3**:29-35

[63] Jongdee B, Fukai S, Cooper M. Leaf water potential and osmotic adjustment as physiological traits to improve drought tolerance in rice. Field Crops Research. 2002;**76**:153-163

[64] Sanders GJ, Arndt SK. Osmotic adjustment under drought conditions. In: Plant Responses to Drought Stress. Springer, Berlin, Germany; 2012. pp. 199-229

[65] Turner DW. An index of osmotic adjustment that allows comparison of its magnitude across species and experiments. Physiologia Plantarum. 2006;**127**:478-482

[66] Smirnoff N. The role of active oxygen in the response of plants to water deficit and desiccation. New Phytologist. 1993;**125**:27-58

[67] Hernandez I, Alegre L, Van Breusegem F, Munne-Bosch S. How relevant are flavonoids as antioxidants in plants? Trends in Plant Science. 2009;**14**:125-132

[68] Hernández-Ortega M, Ortiz-Moreno A, Hernández-Navarro MD, Chamorro-Cevallos G, Dorantes-Alvarez L, Necoechea-Mondragón H. Antioxidant, antinociceptive, and anti-inflammatory effects of carotenoids extracted from dried pepper (*Capsicum annuum* L.). BioMed Research International. 2012;**2012**:524019. pp. 10

[69] Farissi M, Bouizgaren A, Ghoulam C. Antioxidant responses of alfalfa (*Medicago sativa* L.) nodules to water deficit stress. Annals of Experimental Biology. 2014;**2**:16-20

[70] Mouradi M, Bouizgaren A, Farissi M, Ghoulam C. Assessment of deficit irrigation responses of moroccan alfalfa (*Medicago sativa* L.) landraces grown under field conditions. Irrigation and Drainage. In press. 2017;**66**(3)

[71] Zhang B, Li FM, Huang G, Cheng ZY, Zhang Y. Yield performance of spring wheat improved by regulated deficit irrigation in an arid area. Agricultural Water Management. 2006;**79**:28-42

[72] Tejero IG, Zuazo VHD, Bocanegra JAJ, Fernández JLM. Improved water-use efficiency by deficit-irrigation programmes: Implications for saving water in citrus orchards. Scientia Horticulturae. 2011;**128**:274-282

[73] Hanson B, Putnam D, Snyder R. Deficit irrigation of alfalfa as a strategy for providing water for water-short areas. Agricultural Water Management. 2007;**93**:73-80

[74] Savvides A, Ali S, Tester M, Fotopoulos V. Chemical priming of plants against multiple abiotic stresses: Mission possible? Trends in Plant Science. 2016;**21**:329-340

[75] McDonald MB. Seed priming // Seed Technology and Its Biological Basis / Eds. M. Black, J.D. Bewley. Sheffield: Sheffield Academic Press, 2000;287-325

[76] Tavili A, Zare S, Moosavi SA, Enayati A. Effects of seed priming on germination characteristics of bromus species under salt and drought conditions. American-Eurasian Journal of Agricultural and Environmental Science. 2011;**10**(2):163-168

[77] Bhanuprakash K, Yogeesha H. Seed priming for abiotic stress tolerance: An overview. In: Abiotic Stress Physiology of Horticultural Crops. Springer, New Delhi: Springer India; 2016. pp. 103-117

[78] Basra S, Farooq M, Tabassam R, Ahmad N. Physiological and biochemical aspects of pre-sowing seed treatments in fine rice (*Oryza sativa* L.). Seed Science and Technology. 2005;**33**:623-628

[79] Bradford KJ. Manipulation of seed water relations via osmotic priming to improve germination under stress conditions. HortScience (USA). 1986;**21**:1105-1112

[80] Espanany A, Fallah S, Tadayyon A. Seed priming improves seed germination and reduces oxidative stress in black cumin (*Nigella sativa*) in presence of cadmium. Industrial Crops and Products. 2016;**79**:195-204

[81] Kaur S, Gupta AK, Kaur N. Effect of hydro- and osmopriming of chickpea (*Cicer arietinum* L.) seeds on enzymes of sucrose and nitrogen metabolism in nodules. Plant Growth Regulation. 2006;**49**:177-182

[82] Harb E. Effect of soaking seeds in some growth regulators and micronutrients on growth, some chemical constituents and yield of faba bean and cotton plants. Bulletin of Faculty of Agriculture. Cairo Univ. (Egypt). 1992;**43**:429-452

[83] Sadeghi H, Khazaei F, Yari L, Sheidaei S. Effect of seed osmopriming on seed germination behavior and vigor of soybean (*Glycine max* L.). ARPN Journal of Agricultural and Biological Science. 2011;**6**:39-43

[84] Rahimi A. Seed priming improves the germination performance of cumin (*Cuminum cyminum* L.) under temperature and water stress. Industrial Crops and Products. 2013;**42**:454-460

[85] Conrath U. Molecular aspects of defence priming. Trends in Plant Science. 2011;**16**:524-531

[86] Chen K, Arora R. Priming memory invokes seed stress-tolerance. Environmental and Experimental Botany. 2013;**94**:33-45

[87] Conrath U, Beckers GJM, Langenbach CJG, Jaskiewicz MR. Priming for enhanced defense. Annual Review of Phytopathology. 2015;**53**:97-119

[88] Sun YY, Sun YJ, Wang MT, Li XY, Guo X, Hu R, Ma J. Effects of seed priming on germination and seedling growth under water stress in rice. Acta Agronomica Sinica. 2010;**36**:1931-1940

[89] Foyer CH, Noctor G. Redox homeostasis and antioxidant signaling: A metabolic interface between stress perception and physiological responses. The Plant Cell. 2005;**17**:1866-1875

[90] Wang WB, Kim YH, Lee HS, Kim KY, Deng XP, Kwak SS. Analysis of antioxidant enzyme activity during germination of alfalfa under salt and drought stresses. Plant Physiology and Biochemistry: PPB/Societe francaise de physiologie vegetale. 2009;**47**:570-577

[91] Kibinza S, Bazin J, Bailly C, Farrant JM, Corbineau F, El-Maarouf-Bouteau H. Catalase is a key enzyme in seed recovery from ageing during priming. Plant Science: An International Journal of Experimental Plant Biology. 2011;**181**:309-315

[92] Chen K, Arora R, Arora U. Osmopriming of spinach (*Spinacia oleracea* L. cv. Bloomsdale) seeds and germination performance under temperature and water stress. Seed Science and Technology. 2010;**38**:36-48

[93] Mouradi M, Bouizgaren A, Farissi M, Latrach L, Qaddoury A, Ghoulam C. Seed osmopriming improves plant growth, nodulation, chlorophyll fluorescence and nutrient uptake in alfalfa (*Medicago sativa* L.)–rhizobia symbiosis under drought stress. Scientia Horticulturae. 2016;**213**:232-242

[94] Umair A, Ali S, Hayat R, Ansar M, Tareen MJ. Evaluation of seed priming in mung bean (*Vigna radiata*) for yield, nodulation and biological nitrogen fixation under rainfed conditions. African Journal of Biotechnology. 2011;**10**:18122-18129

[95] Rashid A, Harris D, Hollington P, Ali S. On-farm seed priming reduces yield losses of mungbean (*Vigna radiata*) associated with mungbean yellow mosaic virus in the North West Frontier Province of Pakistan. Crop Protection. 2004;**23**:1119-1124

[96] Harris D, Breese WA, Rao JVDKK. The improvement of crop yield in marginal environments using 'on-farm' seed priming: Nodulation, nitrogen fixation and disease resistance. Australian Journal of Agricultural Research. 2005;**56**:1211

[97] Zhang F, Yu J, Johnston CR, Wang Y, Zhu K, Lu F, Zhang Z, Zou J. Seed priming with polyethylene glycol induces physiological changes in sorghum (*Sorghum bicolor* L. Moench) seedlings under suboptimal soil moisture environments. PloS One. 2015;**10**:e0140620

[98] Ansari O, Chogazardi H, Sharifzadeh F, Nazarli H. Seed reserve utilization and seedling growth of treated seeds of mountain rye (*Secale montanum*) as affected by drought stress. Cercetari Agronomice in Moldova. 2012;**45**:43-48

Genetic Variability of US and Czech *Phalaris Arundinacea* L. Wild and Cultivated Populations

Tereza Kávová, Barbora Kubátová,
Vladislav Čurn and Neil O. Anderson

Additional information is available at the end of the chapter

http://dx.doi.org/10.5772/intechopen.69669

Abstract

The spread of invasive plant species in natural habitats has become a worldwide problem with negative impacts. *Phalaris arundinacea*, an important forage and ornamental crop, is widespread worldwide. In recent years there has been a massive spread of *P. arundinacea* across North America and Canada. Production of *Phalaris* seed as a forage crop occurs in northern Minnesota; seeds are sold throughout the world, particularly in central Europe. We tested genetic similarities and differences between populations in the US (Minnesota) and the Czech Republic using ISSRs to determine potential gene flow for this forage crop. The cultivated forage and wild genotypes were dispersed into two groups that overlapped. At least four sets of wild US genotypes are dissimilar to European counterparts and potentially native to N. America. Future work to prove the ancestry of each accession will be necessary. Nonetheless, the sale of forage cultivars related to or derived from European types causes genetic mixing with N. American types. Part of this intercontinental gene flow is exacerbated by the production of *Phalaris* forage seed in Minnesota. The implications of these findings for management of invasive crops native to both continents are significant for forage producers, managers, and breeders.

Keywords: reed canarygrass, invasive species, forage cultivars, ornamental cultivars, ISSRs

1. Introduction

Phalaris arundinacea L., reed canarygrass, is widespread throughout the world, except Antarctica and Greenland [1]. The center of diversity for this genus is in the Mediterranean area; *Phalaris* occur in moist habitats from lower to alpine altitudes. About 22 *Phalaris* species are found mainly in temperate zones of Europe, N. America and South Africa. Among the most important species of *Phalaris* are: *P. arundinacea*, *P. aquatica*, *P. canariensis*, *P. amethystina*, *P. angusta*, *P. brachystachys*, and *P. minor* [1].

Phalaris arundinacea is a wind-pollinated, wetland grass cultivated as a forage and ornamental crop in temperate regions, widely used for soil stabilization, remediation and, more recently, for bioenergy [2–6]. Soil and water restoration projects have also used *P. arundinacea* for phytoremediation [7–10]. Wastewater treatment facilities employ *P. arundinacea* for removal of N [11–13].

Phalaris is widely cultivated both for forage and for ornamental (gardening) purposes. While long cultivated for forage in the US [14] and Sweden [15], its domestication has been relatively recent. Of greater significance is the breeding and cultivation of reed canarygrass in the US and Europe. Existing forage and ornamental cultivars resulted from as few as 1–2 selections and sexual recombination cycles removed from wild types [16], such as 'Chrastava', domesticated in the Czech Republic [17]. 'Chrastava' was genetically similar to wild Czech populations while all other ornamental cultivars differed [18]. High levels of seed dormancy, seed shattering, and low yield potential exist in most populations, e.g. Norwegian [19] but not French [20]. Thus, early forage production trials involved clonal transplants (rhizomes) in Connecticut (in 1834) and New Hampshire (in 1835) [21, 22]. Current seed production within the US occurs in Roseau, Minnesota, which is surrounded by wet meadows.

In Europe, the standard forage cultivar is 'Palaton' (from the US), while other important ones include: 'Luba son. Motycka' (Poland); 'Motterwizer' (Denmark); 'Peti', 'Szarvasi 50', 'Szarvasi 60', 'Keszthelyi 52' (Hungary); 'Lara' (Norway); 'Vantage', 'Venture' (US); 'Bellevue', 'Rival' (Canada); 'Chrastava' (Czech Republic) [23]. Current breeding is focused on improving of its yield potential as a fodder crop as well as for wastewater treatment plants and, more recently, biomass production. *Phalaris* is often used in gardening and ornamental horticulture [18]. It is cultivated mainly as decorative plants with longitudinal white or yellow variegated cultivars from the group *Phalaris arundinacea* var. *picta* and *luteopicta* [24, 25].

Phalaris arundinacea has high dry matter yield (8–12 t ha^{-1}) for forage as well as drought and flood tolerance when compared to timothy (*Phleum pratense*) and tall fescue (*Bromus inermis*) [26, 27]. Three forage cultivars ('Palaton', 'Vantage', 'Venture') responded to selection for establishment capacity with annual weeds [16]. Invasive genotypes possess wide genotype × environment (G×E) interactions across environments for emergence, tiller production, leaf number and biomass, indicating a lack of stability and wide genetic variation [28–30]. Recent molecular studies have shown that central European (Czech) wild populations were genetically similar to the forage 'Chrastava' while differing significantly from ornamental cultivars [18]. In contrast, within MN populations, forage/ornamentals were genetically similar to wild types [31].

Despite unverified assertions that "reed canarygrass is native to the northern half of the United States..." and "native to the temperate portions of Europe, Asia, and North America" [32], invasion biologists and ecologists have consistently postulated that *P. arundinacea* was native to Eurasia but introduced in N. America [33]. Untested hypotheses for *P. arundinacea* invasion in N. America [2, 34] encompass introduction of cultivated types from Eurasia [35], hybridization of Eurasian and N. American populations [28], and/or release of competitive hybrids from breeding programs [36]. However, native N. American *P. arundinacea* populations have been discovered in Ontario, Canada [35] and remote areas elsewhere [37]; herbarium specimens collected in 1825 resembled diploid *P. arundinacea* subsp. *rotgesii* [36]. Recent molecular genetic analyses of herbarium specimens have confirmed the existence of native N. American populations across the continent [38–40]. Nelson et al. [30] determined that the population genetic structure of wild, forage, and ornamental exotic and N. American *Phalaris* harbored a high amount of genetic diversity within, as opposed to among, populations. Thus, range expansion of *P. arundinacea* in N. America is not a result of hybridization among exotic, forage, and native genotypes [38] despite previous theories [28].

Original and introduced *P. arundinacea* populations coexisted in North America for more than a hundred years. We presume that there has been a myriad of migration and intraspecific crossing of this species. It is assumed that the European species and their hybrids are more aggressive [21, 41]. Casler et al. [42] investigated the genetic differences between European and North American genotypes. They found that, on the basis of nuclear DNA, genotypes can be divided into two distinct groups: group one consisted of three closely related genotypes from North America and a group two consisting from other assessed genotypes. Genotypes of first group from Oregon ('Superior'), Alabama ('Auburn') and Arkansas ('AR Upland') could be the sources of the original North American gene pool [42]. These genotypes significantly differed from all European genotypes and it supports the suggestion of their different origins. Casler et al. [42] found ample support for the action of the founder effect resulting from the migration of *Phalaris* from Europe or Asia in recent interglacial periods. These genotypes are, therefore, considered as originating in North America. The founding population in North America, therefore, probably has undergone many mutations that led to the creation genotypes different from Europe. These mutations had little effect on plant morphology and fitness-plant phenotypes remains completely unchanged. As a result, their lower genetic variability results in a bottleneck effect [42].

Previous work by our labs [18, 31] analyzed phenotypic and genotypic markers in genotypes obtained from wild populations growing along the six main rivers within the Czech Republic (Berounka, Dyje, Labe, Lužnice, Orlice, Vltava) and commercial cultivars (forage, ornamental types) grown in the Czech Republic to serve as a foundation for Central European reed canarygrass diversity. ISSRs or inter-simple sequence repeats, for the first time ever, showed distinct genetic differences between ornamental cultivars and wild *P. arundinacea* [18]. Interestingly, the Czech forage and biomass cultivar, 'Chrastava', could not be differentiated from the same wild populations. Most of the genetic diversity was within, rather than among, wild Czech populations [18].

The objective of the present study was to extend the focus on assessment of genetic structure to wild *Phalaris* populations collected in Minnesota (US) along the major rivers and wet meadows or fields with a larger sampling of comparative N. American forage cultivars. Since *Phalaris* seed for forage is commercially produced in Roseau, Minnesota for sale worldwide, sampling in and around production fields in Roseau is part of this study. First, the Minnesota genotypes along with forage comparisons from throughout North America were assessed for genotypic and population differences using ISSRs. Second we analyzed both the Minnesota and Czech [18] genotypic data together to compare differences among continents for genetic structural similarities and differences.

2. Materials and methods

2.1. Genotypes

A total of 16 wild *P. arundinacea* populations were collected in 2012 along the six major rivers in the State of Minnesota, U.S.A. (Des Moines, Minnesota, Mississippi, Red, Roseau, St. Croix) as well as wet meadows or cultivated fields (**Table 1**). The Des Moines, Minnesota, and St. Croix rivers empty southward into the Mississippi river, flowing to the Gulf of Mexico whereas the Roseau and Red rivers flow north into Manitoba, Canada, emptying into Lake Winnipeg. The headwaters for both the Mississippi and Red rivers watersheds originate in Minnesota. Collection protocols for wild *Phalaris* populations followed the same methodology used by Anderson et al. [18] for the Czech populations, with multiple collection sites along each river for a maximum of five genotypes/population (**Table 1**). Seeds of 13 forage cultivars bred, produced and/or grown across North America (**Table 1**) were obtained from the U.S. Department of Agriculture's Germplasm Resources Information Network or USDA-GRIN (http://www.ars-grin.gov/npgs/), germinated and grown to the juvenile stage for harvesting mature leaves. One to five genotypes were analyzed for each accession.

We included data from our previous paper [18] and that of Kávová's M.S. thesis [31] for comparative purposes, namely 110 European genotypes from Czech wild populations (1 site/river; 1–9 genotypes/collection site/river) collected in 2011 along the six main rivers of the

Population or forage cultivar codes	River/wet meadow name and location or forage cultivar name and germplasm source	GPS[1] coordinates for site of collection (wild populations) or germplasm bank identifier number; [citations]
2.1.2	St. Croix River, South of Bayport, MN; by the Bayport Marina	Lat.[2] 45°0′32.8710″ N Long.[2] −92°46′40.4286″ W
6.1.3	St. Croix River, St. Croix State Park; along the river by boat launch and swimming areas	Lat. 45°57.012′ N Long. −92°34.044′ W
8.I.A.1; 8.I.C.3; 8.I.G.3; 8.II.A.2; 8.II.F.3	Wet Meadow, Chanhassen, MN; Horticulture Research Center's "Rice Paddy" wetlands	Lat. 44°51′43.3296″ N Long. −93°35′59.4126″ W
9.3.1	Mississippi River, Reno, MN; along the dead arms, S from the dam of the "big" lake	Lat. 43°36.128′ N Long. 91°16.151′ W
14.2.1	Mississippi River, Red Wing, MN; along the river banks in a wooded area	Lat. 44°35′03.9444″ N Long. 92°38′39.6918″ W
21.5.1	Mississippi River, between Little Falls and Rice, MN; in open areas between wooded banks	Lat. 45°49.597′ N Long. 94°21.262′ W
34.3.1	Mississippi River, near the headwaters; W of Bear Den Landing, Mississippi Headwaters State Forest	Lat. 47°26.012′ N Long. 95°07.748′ W
38.1.B.3	Minnesota River, Blakeley, MN; W of Belle Plaine. MN; in open wet meadows	Lat. 44°36′47.1708″ N Long. 93°51′32.8320″ W
38.2.3	Minnesota River, Blakeley, MN; W of Belle Plaine. MN; in open wet meadows	Lat. 44°36′43.7214″ N Long. 93°51′35.2620″ W

Population or forage cultivar codes	River/wet meadow name and location or forage cultivar name and germplasm source	GPS[1] coordinates for site of collection (wild populations) or germplasm bank identifier number; [citations]
46.1.1	Minnesota River, SE of Montevideo, MN at the confluence of Highways 212/15	Lat. N 44°54′09.8″ N Long. 95°41′07.9″ W
50.1.1	Des Moines River, S of Petersburg, MN at the border with the State of Iowa	Lat. 43°31′33.2″ N Long. 94°55′07.4″ W
54.3.2	Des Moines River, SW of Dovray, MN; adjacent to Highway 8	Lat. 44°00′09.1″ N Long. 95°35′00.3″ W
56.2.2	Roseau River, in the Red Lake State Wildlife Mgt. Area, W of Mulligan Lake, adjacent to the Red Lake Indian Reservation; Co. Rd. 704, at headwaters (source) of the river	Lat. 48°32.774′ N Long. 95°19.204′ W
58.1.3	Roseau River, N of Roseau, MN; wet meadows near Hwy. 3	Lat. 48°54.504′ N Long. 95°49.778′ W
58.2.2	Roseau River, N of Roseau, MN; wet meadows near Hwy. 3	Lat. 48°54.546′ N Long. 95°49.711′ W
58.3.1	Roseau River, N of Roseau, MN; wet meadows near Hwy. 3	Lat. 48°54.562′ N Long. 95°49.635′ W
58.IV.A.1	Roseau River, N of Roseau, MN; wet meadows near Hwy. 3; transect in cultivated field	Lat. 48°54.699′ N Long. 95°52.130′ W
58.IV.H.3	Roseau River, N of Roseau, MN; wet meadows near Hwy. 3; transect in cultivated field	Lat. 48°54.753′ N Long. 95°52.084′ W
61.1.2	Roseau River, Caribou, MN; Hwy. 4 near confluence with State Ditch; S of the Canadian Border	Lat. 48°59.006′ N Long. 96°26.951′ W
63.4.3	Red River, S of McCauleville, MN and SW of Kent, MN	Lat. 46°26′43.0″ N Long. 96°42′57.9″ W
74.1.2	Red River, Oslo, MN; S of Big Woods, County Ditch 38	Lat. 48°18′40.3″ N Long. 97°07′24.4″ W
VEN	'Venture' (Minnesota); derived from crossing 'Vantage' × 'Flare'); low alkaloid variety; does not contain any tryptamine-carboline alkaloids; USDA-GRIN; https://npgsweb.ars-grin.gov	PI[3] 531089 [14, 42]
PAL	'Palaton' (Minnesota); derived from 'Flare', 'Vantage' and 'Rise'); low alkaloid variety; does not contain any tryptamine-carboline alkaloids; USDA-GRIN; https://npgsweb.ars-grin.gov	PI 531088 [14]
AUB	'Auburn' (Alabama); landrace, most likely derived from native N. American germplasm; USDA-GRIN; https://npgsweb.ars-grin.gov	PI 422031 [42]
IOR	'Ioreed' (Iowa); high levels of alkaloids; USDA-GRIN; https://npgsweb.ars-grin.gov	PI 422030 [42]
365	367 (British Columbia, Canada); USDA-GRIN; https://npgsweb.ars-grin.gov	PI 387929

Population or forage cultivar codes	River/wet meadow name and location or forage cultivar name and germplasm source	GPS[1] coordinates for site of collection (wild populations) or germplasm bank identifier number; [citations]
PHA	*Phalaris arundinacea*; USDA-GRIN; https://npgsweb.ars-grin.gov	PI 241065
PN-609	Unknown origin; USDA-GRIN; https://npgsweb.ars-grin.gov	PI 371754
GRO	'Grove' (Ontario, Canada); USDA-GRIN; https://npgsweb.ars-grin.gov	PI 357645 [42]
MN-76	MN-76 (Minnesota) 4-clone double cross hybrid; low alkaloid variety; does not contain any tryptamine-carboline alkaloids; USDA-GRIN; https://npgsweb.ars-grin.gov	PI 578797 [42]
CANA	'Cana' (California); USDA-GRIN; https://npgsweb.ars-grin.gov	PI 578795
VAN	'Vantage' (Iowa); high alkaloid content; USDA-GRIN; https://npgsweb.ars-grin.gov	PI 578794 [14, 42]
MCRC1	NCRC-1 (Minnesota); USDA-GRIN; https://npgsweb.ars-grin.gov	PI 578793
SUP	'Superior' (Oregon); most likely derived from native N. American germplasm; USDA-GRIN; https://npgsweb.ars-grin.gov	PI 578792 [14, 42]

[1]GPS, global positioning system.

[2]Lat., latitude; Long., longitude.

[3]PI, plant introduction; USDA-GRIN, U.S. Dept. of Agriculture, Germplasm Resources Information Network.

Table 1. Minnesota (U.S.A.) population or North American forage cultivar codes, river/wet meadow name and location or forage cultivar name and germplasm source; GPS coordinates for site of collection (wild populations) or germplasm bank identifier number for *Phalaris arundinacea* wild populations collected in the State of Minnesota (MN; U.S.A.) along rivers and in wet meadows.

Czech Republic (Berounka, Dyje, Labe, Lužnice, Orlice, and Vltava). Similar to the Minnesota wild populations, five of the Czech rivers empty into the North Sea basin while the Dyje River flow into the Black Sea basin [18]. Additional wild population samples were made at the OSEVA PRO, Ltd., Grassland Research Station (Rožnov-Zubří, CZ); commercial forage, ornamental cultivars either bred and/or grown in the Czech Republic were also included. These all were grown and previously analyzed in our previous study [18] and ISSR molecular data from these were used herein to compare with the results found with the Minnesota and N. American types. Genotypic codes for all Czech germplasm consisted of the following: BE-1, 2, 3 (Berounka); DY1, 2, 3 (Dyje); LA1 (Labe); LU1, 2, 3 (Lužnice); OR1, 3 (Orlice); VL1, 3 (Vltava); CHR ('Chrastava'; forage cultivar); Z13, Z77, Z83, Z124, Z125 (OSEVA PRO, Ltd, Grassland Research Station, Rožnov-Zubří, CZ); ZP/COV1, 17 (Gardening Pelikán, Spálené Poříčí), AT/P6, 7 ('Picta'), AT/T2, 6 ('Tricolor'), F/L1, 4 ('Luteopicta'), F/Pa3, 4 (*Phalaris arundinacea*), F/P2, 3 ('Picta'), SF/P4, 5 ('Picta'). Any clonal ramets of genotypes were coded alphabetically (A, B, C, etc.) at the end of the genotypic code.

2.2. Genetic analyses

Genetic variability was assessed using ISSR markers. This molecular technique is often used in studies focused on genetic variation of plant populations and plant germplasm and we verified its suitability and stability in analyses of Phalaris genotypes. ISSR is also marker system with high detectable extent of genetic variation/diversity and also with the ability to detect the genetic diversity among individual accessions.

2.3. DNA extraction and ISSR analyses

DNA extraction from leaf samples and subsequent ISSR analyses of all Minnesota and N. American samples followed the protocols delineated by Kávová [31]. Four primers from the University of British Columbia were used to generate scorable ISSR markers: UBC 810—$[GA]_8T$, UBC 825—$[AC]_8T$, UBC 881—$G3[TGGGG]_2TG$, and UBC 890—$VHV[GT]_7$ [31]; these have been used in our subsequent studies for *Phalaris* [18, 30, 43]. Seventy-six markers (MW = 270–1200 base pairs [bp]) were scored and transformed into a binary character matrix (1 = present, 0 = absent).

2.4. Statistical analyses

Genetic distance matrices were created with Nei and Li's [44] metrics. PCoA (principal coordinate analysis) and UPGMA (unweighted pair group method with arithmetic mean) cluster analyses were calculated with MVSP, version 3.1 (Multi-Variate Statistical Package; Kovach Computing Services U.K.) and DARwin, version 5.0.158 (Dissimilarity Analysis and Representation for windows; CIRAD, F) software. Genetic structure was calculated using STRUCTURE version 2.3.4, a Bayesian clustering algorithm (Admixture Model; correlated allele frequencies; K = 2, K = 4, K = 6, and K = 10 groupings; 100,000 burnin repetitions) [43, 45]. STRUCTURE groupings refer to relationship patterns. After plotting, the K = 2 grouping had the necessary decrease in slope and increase in variance, diagnostic of the true K value, with the greatest number of genotypes/grouping; all other groupings were eliminated [30]. Only results from the K = 2 grouping will be shown.

3. Results

The four ISSR primers generated 76 scorable bands (56.6% were polymorphic). The UPGMA cluster analysis showed three distinct grouping of genotypes, all of which separated at a genetic distance of 0.0 (**Figure 1**). The first grouping consisted of strictly forage cultivars from Iowa (PAL, VEN), Minnesota (MN-76), California (CANA) and Missouri (AUB) (**Table 1**), all of which differed significantly ($p \leq 0.05$) from other forage cultivars and wild populations. The next grouping had 4 wild populations from the Mississippi (34.3.1, 38.2.3), Minnesota (46.1.1), Red (63.4.3) rivers in one small grouping, along with another grouping. This latter grouping was subdivided into (a) 7 wild populations from the wet meadow in Chanhassen (8.II.F.3), the Roseau (56.2.2, 58.IV.H.3; 61.1.2), St. Croix (2.1.2; 6.1.3), and Des Moines (50.1.1) rivers and (b)

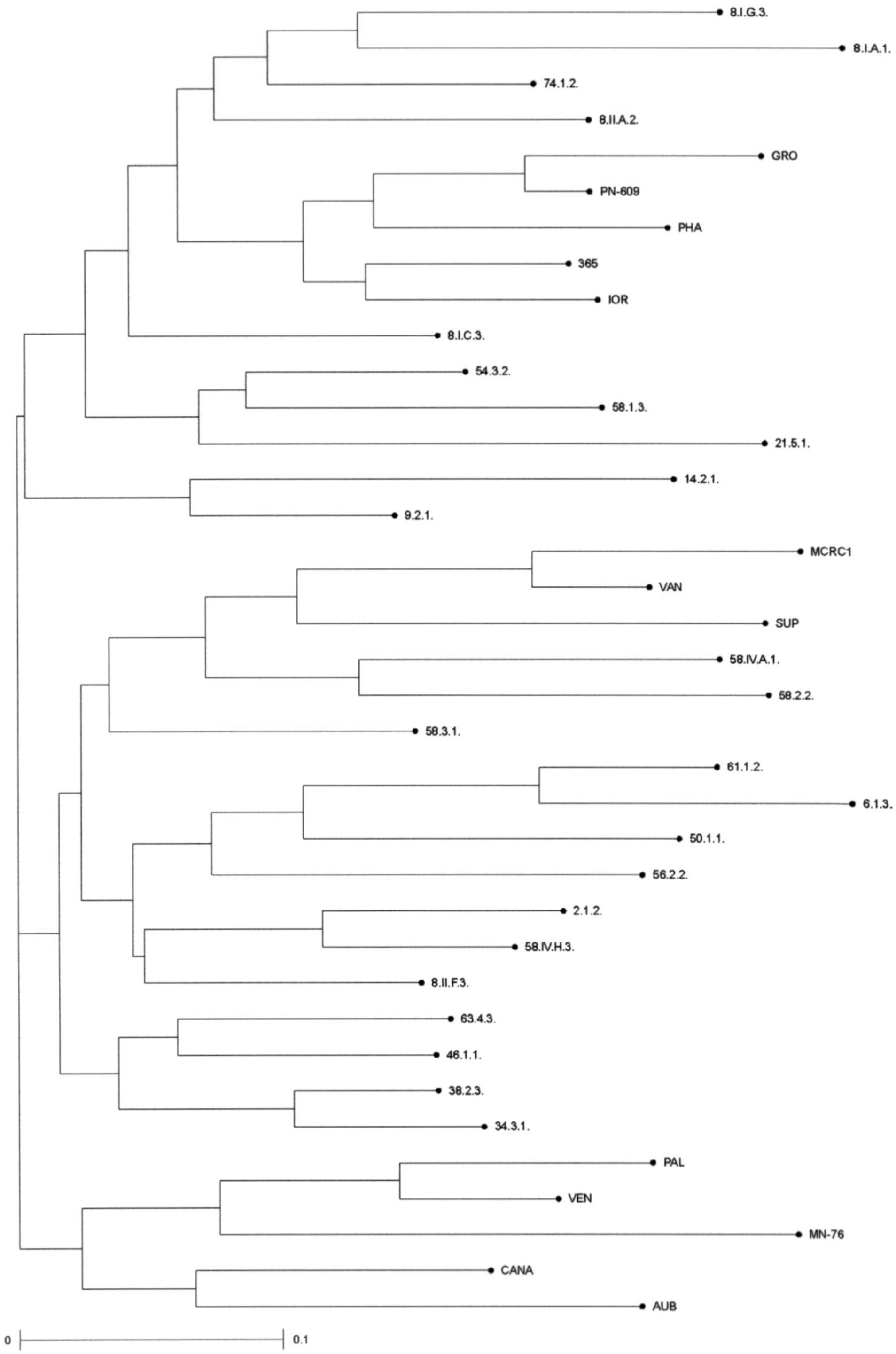

Figure 1. UPGMA, based on ISSR markers, for wild Minnesota populations and N. American comparative forage cultivars of *Phalaris arundinacea*. See **Table 1** for genotypic codes.

http://dx.doi.org/10.5772/intechopen.69669

1 wild population from the Roseau river (58.3.1, 58.2.2, 58.IV.A.1) plus three forage cultivars from Oregon (SUP), Iowa (VAN) and Minnesota (MCRC1) (**Figure 1**). The final grouping consisted of two major subgroupings with (a) 2 wild populations from the Mississippi river (9.2.1; 14.2.1) and (b) a quadriplex set of (i) 3 wild populations from the Mississippi (21.5.1), Roseau (58.1.3), and Des Moines (54.3.2) rivers; (ii) the wet meadow in Chanhassen (8.I.C.3); (iii) 4 forage cultivars from Missouri (IOR), unnamed (PHA), unknown (PN-609), and Ontario, Canada (GRO); (iv) 2 wild populations from the wet meadow in Chanhassen (8.II.A.2; 8.I.A.1; 8.I.G.3) (**Figure 1**).

Principal coordinate analysis (PCoA) of inter-simple sequence repeat (ISSR) markers in reed canarygrass samples from Minnesota and N. America showed two overlapping groupings for the forage and wild genotypes (**Figure 2**). The forage cultivars AUB, VEN, PAL, and MN-76 were the farthest away from the wild populations collected along Minnesota rivers and in wet meadows or fields (**Figure 2**). Other forage cultivars (IOR, PHA, GRO, PN-609, VAN; **Figure 2**) were also categorically and genetically similar to these but more closely related to the wild genotypes.

When the wild and cultivated US genotypes were comparatively analyzed for PCoA together with the Czech/European genotypes [2] this resulted into forming two primary clusters (**Figure 3**). Cluster I (lower circle) included all samples from wild Czech (European) populations along rivers and the forage 'Chrastava' as established for European genotypes by Anderson et al. [18]; this cluster was enriched with all samples of US origin. All US genotypes

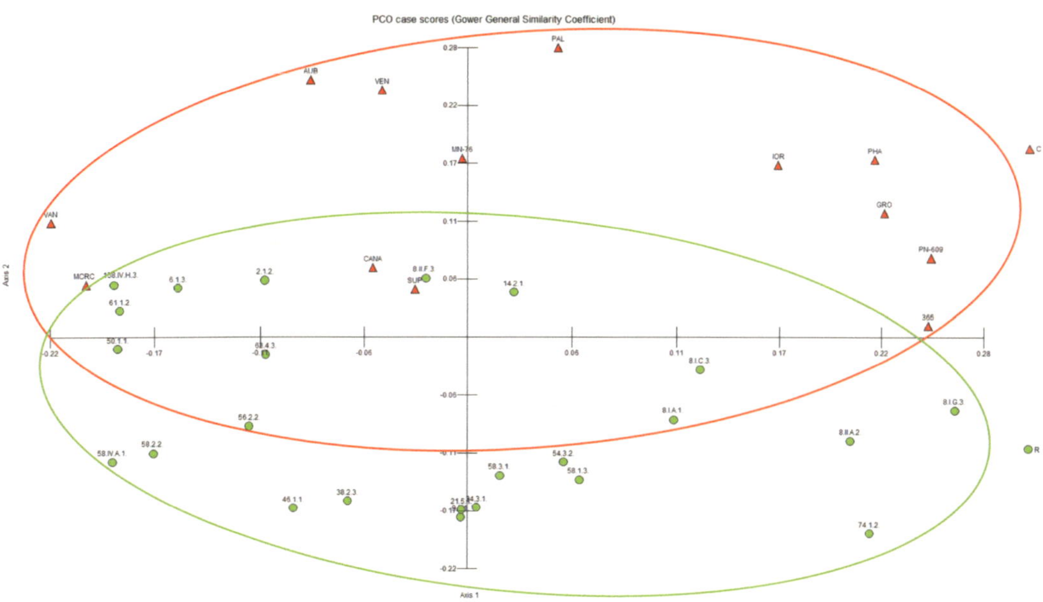

Figure 2. Principal coordinate analysis (PCoA) of inter-simple sequence repeat (ISSR) markers in reed canarygrass samples from US (Minnesota) and N. America. For genotype codes, refer to **Table 1**. Keys to symbols are: cultivated (triangles), wild (circles). The upper oval encompasses the majority of cultivated samples while the lower oval surrounds predominantly wild types.

were clustered into a small, oval sub cluster of Cluster I, on the border of the European wild genotypes and showing high similarity in ISSR marker pattern (**Figure 3**). Cluster II (upper oval) is represented by European horticultural and forage cultivars and genotypes from The Nursery of Genetic Resources, OSEVA PRO, Ltd., Grassland Research Station (Rožnov-Zubří, Czech Republic) with both variegated and nonvariegated leaf types [18].

Assessing the genetic structure of analyzed Czech cultivated and wild genotypes showed classification of genotypes according to Q1/Q2 values (membership probabilities in the C [rows or genotypes] × K [columns or clusters] matrix for a single cluster analysis); K = 2 had the best stratification in STRUCTURE (**Figure 4**). One group, 'PN-609', contains several forage cultivars and a few wild genotypes from Site 8. Whereas the larger group, '54.3.2' contains the remaining genotypes from all rivers, wet meadows and any remaining forage cultivars.

UPGMA analyses of both the US and Czech populations, based on ISSRs, showed distinct groupings of reed canarygrass genotypes (**Figure 5**). The first group was a small set of 6 genotypes, ZPCOV, collected at The Nursery of Genetic Resources, OSEVA PRO, Ltd., Grassland Research Station (Rožnov-Zubří, Czech Republic). The second grouping was a large series of sub clusters divided as follows. The most distant genotypes from the ZPCOV cluster were primarily horticultural cultivars from the Czech Republic along with one sole US genotype from the wet meadow in Chanhassen, MN (8.I.A.1; **Figure 5** and **Table 1**). The next cluster

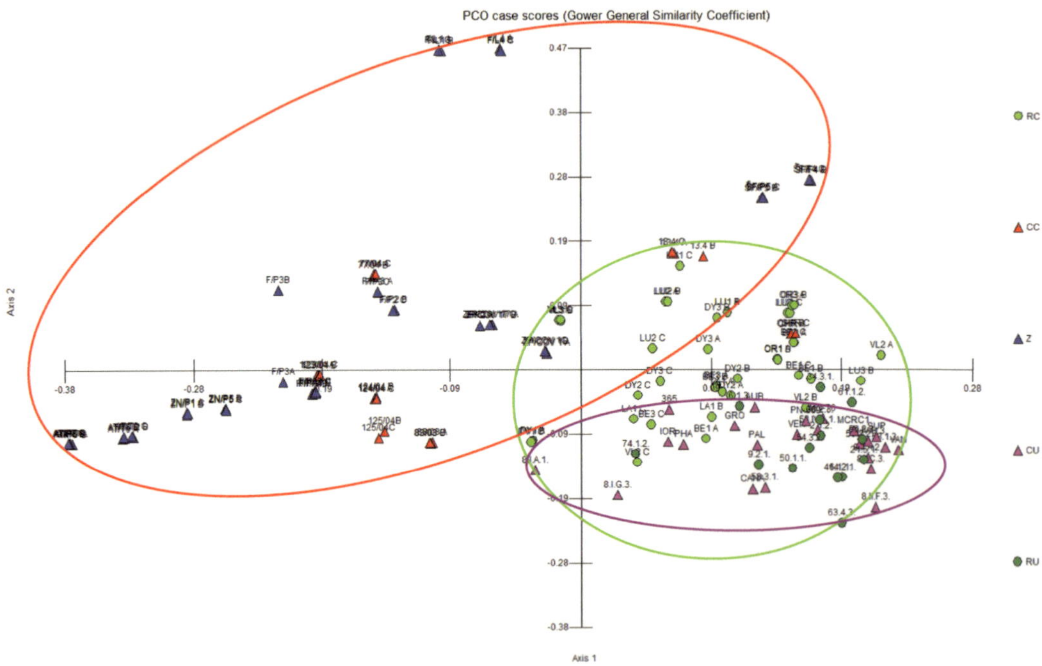

Figure 3. Principal coordinate analysis (PCoA) of inter-simple sequence repeat (ISSR) markers in reed canarygrass samples from Minnesota and N. America compared with Czech wild populations and cultivars [18]. For Minnesota and N. American genotype codes, refer to **Table 1**; for the Czech genotypes, refer to [18] (*cf.* **Table 1**). Keys to symbols are: cultivated (triangles), wild (circles). The large oval (Cluster II) encompasses the majority of cultivated samples while the circle (Cluster I) surrounds predominantly wild types. Key: RC—rivers CZ, RU—rivers MN, CC—CZ forage cultivars, CU—MN forage cultivars, Z—horticultural genotypes.

A

Sample	Type	Q1	Q2
PN-609	C	0.022	0.978
365	C	0.024	0.976
PHA	C	0.024	0.976
IOR	C	0.029	0.971
GRO	C	0.030	0.970
8.I.G.3.	W	0.032	0.968
8.II.A.2.	W	0.096	0.904
PAL	C	0.126	0.874
74.1.2.	W	0.159	0.841
8.I.C.3.	W	0.260	0.740
8.I.A.1.	W	0.347	0.653
MN-76	C	0.389	0.611
VEN	C	0.428	0.572
54.3.2.	W	0.563	0.437
14.2.1.	W	0.572	0.428
58.1.3.	W	0.573	0.427
8.II.F.3.	W	0.648	0.352
AUB	C	0.650	0.350
CANA	C	0.675	0.325
21.5.1.	W	0.692	0.308
58.3.1.	W	0.723	0.277
SUP	C	0.759	0.241
34.3.1.	W	0.808	0.192
9.2.1.	W	0.828	0.172
2.1.2.	W	0.886	0.114
6.1.3.	W	0.936	0.064
56.2.2.	W	0.936	0.064
63.4.3.	W	0.947	0.053
38.2.3.	W	0.949	0.051
46.1.1.	W	0.951	0.049
50.1.1.	W	0.970	0.030
58.IV.A.1.	W	0.971	0.029
61.1.2.	W	0.972	0.028
MCRC1	C	0.973	0.027
58.IV.H.3.	W	0.974	0.026
VAN	C	0.975	0.025
58.2.2.	W	0.976	0.024

B

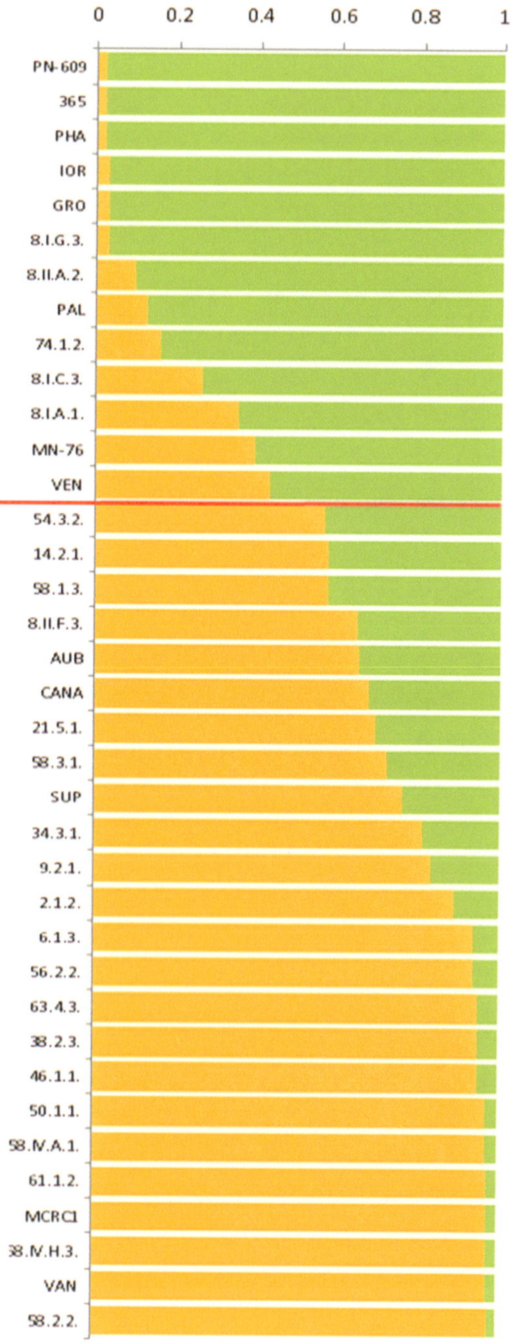

Figure 4. Genetic structure analysis of the US and N. American reed canarygrass collection using STRUCTURE software package (Admixture Model, allele frequencies correlated, K = 6, length of burnin period: 100,000). Key: the population code is located left from the corresponding color bars with two groups of accessions: black— 'PN-609'; grey—'54.3.2'.

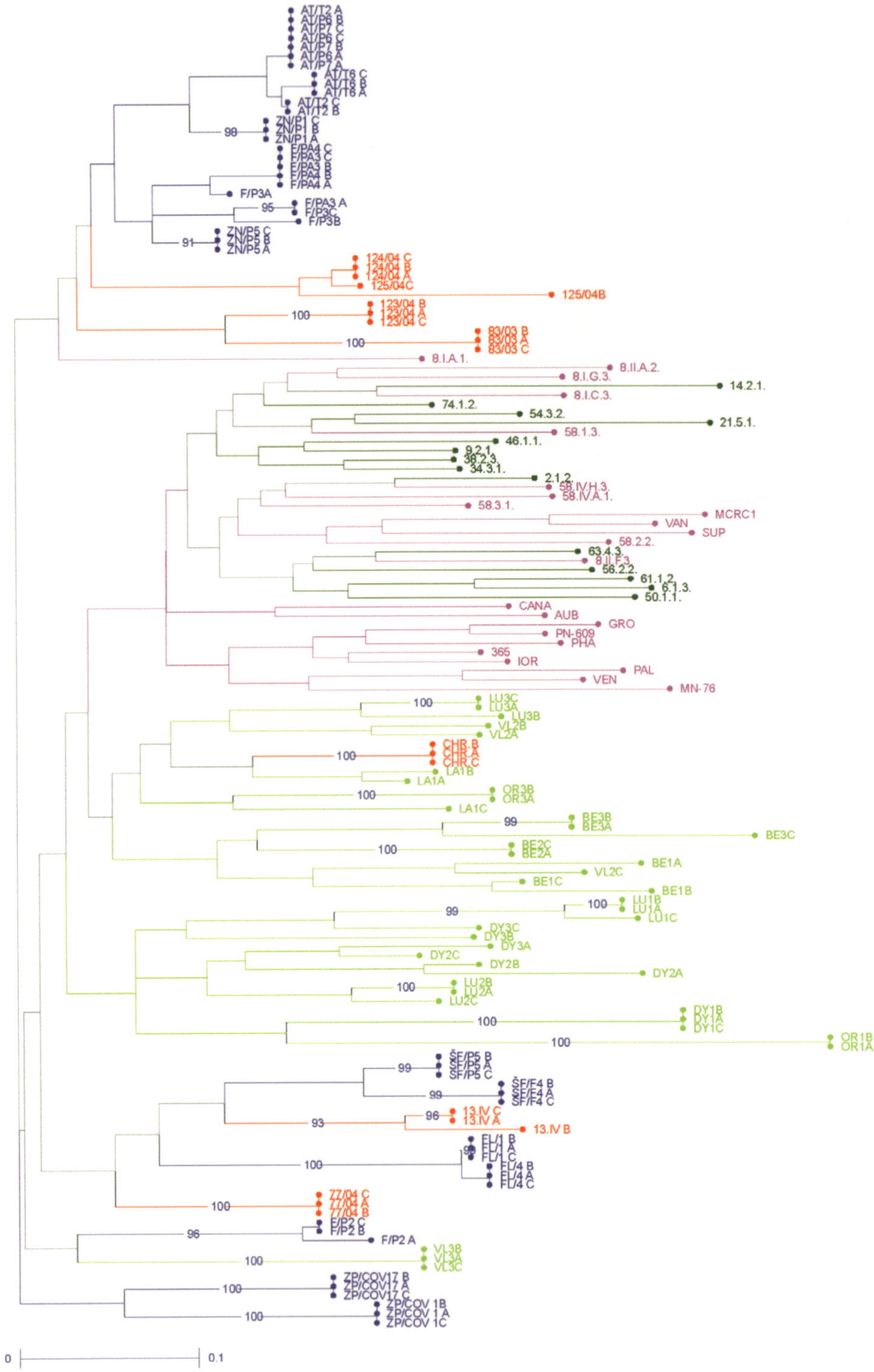

Figure 5. UPGMA analysis, based on ISSR data, of combined Czech, Minnesota and N. American reed canarygrass samples analyzed (*cf.* **Table 1** [18] for genotypic codes of US genotypes for CZ genotypes).

was divided into two groups of: (a) 6 Czech genotypes (3 wild from the Vltava River and 3 GFP/GNP or 'Picta'). Next were two sub clusters which bifurcated at a genetic distance of ~ 0.2 (**Figure 5**). One formed a small grouping of 18 genotypes, namely Czech accessions and MN genotypes while the other was a large grouping of all remaining wild and cultivated US and CZ genotypes.

4. Discussion

There were two overlapping groupings for the forage cultivars and wild reed canarygrass samples from Minnesota and N. America (**Figure 2**). The forage cultivars AUB, VEN, PAL, and MN-76 were the farthest away from the wild populations collected along Minnesota rivers and in wet meadows or fields (**Figure 2**). This included at least one forage cultivar, AUB ('Auburn'), which is most likely derived from native N. American strains (**Table 1**) [7]. SUP ('Superior'), also derived from native N. American strains, was more closely aligned with some wild genotypes, particularly one from the wet meadow in Chanhassen, MN (8.II.F.3; **Table 1**). Other forage cultivars (IOR, PHA, GRO, PN-609, VAN; **Figure 2**) were also categorically and genetically similar to these but more closely related to the wild genotypes.

Based on the UPGMA analysis of US cultivated and wild types of reed canarygrass (**Figure 1**), potentially the 4 wild populations from the Mississippi (34.3.1, 38.2.3), Minnesota (46.1.1), and Red (63.4.3) rivers are the least related to the N. American forage cultivars SUP ('Superior'), VAN ('Vantage') and MCRC-1 and may be native American genotypes. These MN wild populations also differed from the Czech wild populations (**Figure 4**). Casler et al. [42] and Jakubowski et al. [38, 39] used 15 SSR molecular markers to distinguish among N. American native and exotic (European) *P. arundinacea* herbaria specimens. They found that the forage cultivars AUB ('Auburn') and SUP ('Superior'), used in the present study, were native American in origin. However, in our study, these two forage cultivars were even further away from the 4 wild populations identified above. Thus, it may be possible that additional native N. American strains included herein exist. Future work will be devoted to identifying this possibility using the 15 SSR markers specific to N. American *Phalaris* already identified [38, 39, 42].

In the STRUCTURE analysis of the US reed canarygrass collected along Minnesota rivers and in wet meadows, along with the North American cultivars, the cultivars were distributed throughout both groups (**Figure 4A** and **B**). This was unexpected and surprising since, for instance, the Red and Roseau Rivers running through northern Minnesota do not flow to the Gulf of Mexico and the Atlantic Ocean via the Mississippi River, but instead flow to Manitoba, Canada into Lake Winnipeg and would have limited opportunities for gene exchange. Additionally, since reed canarygrass is native in Minnesota, there could have been divergent evolution within isolated rivers creating distinct populations but this was not found to be the case. This could be due to wind pollination, which may allow for gene flow (pollen) between rivers. Also likely could be the small sample sizes collected along all rivers and/or the choice of genetic markers that, even though they are polymorphic among the populations and cultivars, may not be able to discriminate among *Phalaris* along all rivers.

In the PCoA and STRUCTURE analyses of both the N. American and European sample sets from river habitats and forage cultivars, no clear differentiation among groupings was evident (**Figure 5**). The pattern of genetic markers in the European (Czech) genotypes from alluvial habitats was inclusive of all US wild and cultivated forage genotypes. Both groups of genotypes (wild/cultivated) overlapped and, in contrast with our previous analysis of the European genotypes [18], it was not possible to distinguish between wild and cultivated genotypes with precision. One reason for this may be the low genetic variation and differentiation among genotypes and their high genetic similarity. What is surprising is the very low extent of genetic variability among US genotypes, which formed one "dense" group in this pooled analysis, with low levels of genetic dissimilarity. This fact may also explain poor differentiation between US wild genotypes and US cultivated forage, because of their low genetic dissimilarity. Another reason may be the small sample sizes tested herein. Future work will be devoted to conducting a more thorough sampling of wild and commercial *P. arundinacea* throughout Minnesota and analyzing the SSR genetic alignment into the distinct European vs. native N. American haplotypes.

Since all grasses, including reed canarygrass, are anemophilous (wind-pollinated), it would be easy for genetic mixing to occur in adjacent plantings of cultivated and wild types. Likewise, as most forage cultivars bred and/or produced in Minnesota and N. America are closely related to or derived from European types [42], this also could be a reason why the wild and forage types overlapped in their genetic similarity (**Figures 1** and **2**). The numerous influxes of exotic, European types and cross-pollination effects (either occurring naturally or by hand pollination by plant breeders), combined with migration have mixed the gene pools [42]. For instance, while 'Rival' has both European and Scandinavian ancestors, 'Ioreed' is a hybrid mixture with the European nuclear haplotype but N. American cytoplasmic haplotype [42]. However, maintaining the integrity of N. American *Phalaris* germplasm, distinct from the exotic or European forage types commonly distributed on the continent [33], is of paramount importance given its historical and cultural significance in weavings by native Americans [46–49]. Destruction of native *Phalaris* genotypes would violate Treaty Rights.

5. Conclusion

In Minnesota populations of *Phalaris*, the cultivated and wild genotypes formed separate groups, which did overlap significantly. At least four sets of wild U.S. genotypes are the most dissimilar to European counterparts and, as such, could be native to N. America. Future work to prove the ancestry of each accession will be necessary. Nonetheless, the sale of forage cultivars related to or derived from European types continues to cause genetic mixing with N. American types. Part of this intercontinental gene flow and exchange is exacerbated by the production of *Phalaris* forage seed in Minnesota, which is sold both in N. America and Europe. While the expectation that forage/ornamental reed canarygrass cultivars should have a similar genetic makeup with the wild populations across continents (due to limited breeding and genetic selection pressures in this forage and ornamental crop) this is clearly not the case despite *Phalaris* being an invasive, wind-pollinated grass. The implications of these findings for management of invasive crops native to both continents have significant implications for forage producers, managers, and breeders.

Author details

Tereza Kávová[1], Barbora Kubátová[1], Vladislav Čurn[1] and Neil O. Anderson[2]*

*Address all correspondence to: ander044@umn.edu

1 Faculty of Agriculture, Biotechnological Centre, University of South Bohemia, České Budějovice, Czech Republic

2 Department of Horticultural Science, University of Minnesota, St. Paul, MN, USA

References

[1] Baldini RM. Revision of genus *Phalaris arundinacea*. Webbia. 1995;**49**:265-329

[2] Lavergne S, Molofsky J. Reed canary grass (*Phalaris arundinacea*) as a biological model in the study of plant invasions. Critical Reviews in Plant Sciences. 2004;**23**:415-429

[3] Lewandowski I, Scurlock JMO, Lindvall E, Christou M. The development and current status of perennial rhizomatous grasses as energy crops in the US and Europe. Biomass & Bioenergy. 2003;**25**:335-361

[4] Nilsson D, Hansson PA. Influence of various machinery combinations, fuel proportions and storage capacities on cost of co-handling of straw and reed canary grass to district heating plants. Biomass & Bioenergy. 2001;**20**:247-260

[5] Rice J, Pinkerton B. Reed canarygrass survival under cyclic inundation. Journal of Soil and Water Conservation. 1993;**48**:132-135

[6] Sheaffer CC, Rosen CJ, Gupta SC. Reed canarygrass forage yield and nutrient uptake on a year-round wastewater application site. Journal of Agronomy and Crop Science. 2008;**194**:465-469

[7] Chekol T, Vough LR, Chaney RL. 2002. Plant-soil-contaminant specificity affects phytoremediation of organic contaminants. International Journal of Phytoremediation. 2002;**4**:17-26

[8] Lasat MM, Norvell WA, Kochian LV. 1997. Potential for phytoextraction of Cs-137 from a contaminated soil. Plant Soil. 1997;**195**:99-106

[9] Olsen FJ, Chong SK. Reclamation of acid coal refuse. Landscape Urban Planning.1991;**20**: 309-313

[10] Samecka-Cymerman A, Kempers AJ. Concentrations of heavy metals and plant nutrients in water, sediments and aquatic macrophytes of anthropogenic lakes (former open cut brown coal mines) differing in stage of acidification. Science of the Total Environment. 2001;**281**:87-98

[11] Vymazal J. Constructed wetlands for wastewater treatment in the Czech Republic. Water Science & Technology. 2001;**44**:369-374

[12] Vymazal J. Vegetation development in subsurface flow constructed wetlands in the Czech Republic. Ecological Engineering. 2013;**61**:575-581

[13] Zhu T, Sikora FJ. Ammonium and nitrate removal in vegetated and unvegetated gravelbed microcosm wetlands. Water Science & Technology. 1995;**32**:219-228

[14] Sheaffer CC, Marten GC, Rabas DL, Martin NP, Miller DW. Reed Canarygrass. Minnesota Agricultural Experiment Station Bulletin No. 595-1990 (Item No. AD-SB-5533). St. Paul, MN; 1990

[15] Alway FJ. Early trials and use of reed canary grass as a forage plant. Journal. American Society of Agronomy. 1931;**23**:64-66

[16] Casler MD, Undersander DJ. Selection for establishment capacity in reed canary grass. Crop Science. 2006;**46**:1277-1285

[17] Cagaš B. Reed canary grass 'Chrastava'. Czech Journal of Plant Breeding and Genetics. 2008;**44**(1):41-42

[18] Anderson NO, Kávová T, Bastlová D, Čurn V, Kubátová B, Edwards KR, Januš J, and Květ J. Phenotypic and genotypic variation in Czech forage, ornamental and wild populations of reed canarygrass. Crop Science. 2016;**56**(5):2421-2435

[19] Juntila O, Landgraff L, Nilsen AJ. Germination of *Phalaris* seeds: Seed problems. Acta Horticulturae. 1978;**82**:163-166

[20] Conchou O, Pautou G. 1987. Modes of colonization of an heterogenous alluvial area on the edge of the Garonne River by *Phalaris arundinacea*. Regulated Rivers. 1987;**1**:37-48

[21] Anonymous. Ribbon grass. New England Farmer. 1834;**13**:41, 129

[22] Harris A. Ribbon grass. New England Farmer. 1835;**14**:125

[23] Strasil Z. Ekonomická analýza vybraných energetických rostlin urených prospalování. In: Sbor. Technika a technologie pro nepotravináské vyuití pody a jejíudrování v klidu., Brno. 2000; pp. 17-22

[24] Dostál J. Nová kvtena CSSR 2. Prague: Academia Praha; 1989

[25] Weber E. Invasive Plants of the World. CAB International, Wallingford, UK: CABI Publishing; 2003

[26] Čížková H, Rychterová J, Hamadejová L, Suchý K, Filipová M, Květ J, Anderson NO. Biomass production in permanent wet grasslands dominated with Phalaris arundinacea: Case study of the Třeboň basin biosphere reserve, Czech Republic. In: Vymazal V, editor, The Role of Natural and Constructed Wetlands in Nutrient Cycling and Retention on the Landscape. Springer International, Cham (ZG), Switzerland. 2015. DOI 10.1007/978-3-319-08177-9_1

[27] Gyulai G, Mester Z, Kiss J, Szeman L, Idnurm A, Heszky L. Somaclonal breeding of reed canary grass (*Phalaris arundinacea* L.). Grass and Forage Science. 2003;**58**:210-214

[28] Lavergne S, Molofsky J. Increased genetic variation and evolutionary potential drive the success of an invasive grass. Proceedings of the National Academy of Sciences of the United States of America. 2007;**104**:3883-3888

[29] Morrison SL, Molofsky J. Effects of genotypes, soil moisture, and competition on the growth of an invasive grass, *Phalaris arundinacea* (reed canary grass). Canadian Journal of Botany. 1998;**76**:1939-1946

[30] Nelson MF, Anderson NO, Casler MD, Jakubowski AR. Population genetic structure of N. American and European *Phalaris arundinacea* L. as inferred from inter-simple sequence repeat markers. Biological Invasions. 2014;**15**:353-363. DOI: 10.1007/s10530-013-0525-9

[31] Kávová T. Genetická variabilita v populacích chrastice rákosovité (*Phalaris arundinacea* L). MS Thesis. University of South Bohemia, Ceske Budejovice, CZ. 2013

[32] Schoth HA. Reed Canary Grass. Washington, DC: United States Department of Agriculture. Farmers' Bull. No. 1602; 1938

[33] Galatowitsch SM, Anderson NO, Ascher P. Invasiveness in wetland plants in temperate North America. Wetlands. 1999;**19**:733-755

[34] Saltonstall K. Cryptic invasion by a non-native genotype of the common reed, *Phragmites australis*, into North America. Proceedings of the National Academy of Sciences of the United States of America. 2002;**99**:2445-2449

[35] Dore WG, McNeill J. Grasses of Ontario. Ottawa, Canada: Agriculture Canada, Research Branch, Biosystematics Research Institute; 1980

[36] Merigliano MF, Lesica P. The native status of reed canarygrass (*Phalaris arundinacea* L.) in the inland northwest, USA. Natural Areas Journal. 1998;**18**:223-230

[37] Lavoie C, Dufresne C, Delisle FF. The spread of reed canary grass (*Phalaris arundinacea*) in Québec: A spatio-temporal perspective. Écoscience. 2005;**12**(3):355-375

[38] Jakubowski AR, Casler MD, Jackson RD. Has selection for improved agronomic traits made reed canarygrass invasive? PLoS ONE. 2011;**6**(10):e25757. DOI: 10.1371/journal.pone.0025757

[39] Jakubowski AR, Casler MD, Jackson RD. Genetic evidence suggests a widespread distribution of native North American populations of reed canarygrass. Biological Invasions. 2013;**15**:261-268

[40] Jakubowski AR, Jackson RD, Casler MD. The history of reed canarygrass in North America: Persistence of natives among invading Eurasian populations. Crop Science. 2014;**54**(1):210-219

[41] Maurer DA, Zedler JB. Differential invasion of a wetland grass explained by tests of nutrients and light availability on establishment and clonal growth. Oecologia. 2002;**131**: 279-288

[42] Casler MD, Phillips MM, Krohn AL. DNA polymorphisms reveal geographic races of reed canarygrass. Crop Science. 2009;**49**:2139-2148

[43] Nelson MF, Anderson NO. How many marker loci are necessary? Analysis of dominant marker data sets using two popular population genetic algorithms. Ecology and Evolution. 2013;**3**(10):3455-3470

[44] Nei M, Li W-H. 1979. Mathematical model for studying genetic variation in terms of restriction endonucleases. Proceedings of the National Academy of Sciences of the United States of America. 1979;**76**(10):5269-5273

[45] Pritchard JK, Stephens M, Donnelly P. Inference of population structure using multilocus genotype data. Genetics. 2000;**155**:945-959

[46] Densmore F. How Indians Use Wild Plants for Food, Medicine, and Crafts. Courier Dover Pub., Mineola, New York; 1974

[47] Kindscher K, Noguera E. Cultural use of plants from the Baker Wetlands. Prepared for the technical outreach services for native American communities (TOSNAC) at Haskell Indian Nations University. Manhattan: Kansas Biological Survey, University of Kansas; 2002

[48] Lerman N. Okanogan (Salish) ethnology. Field notes and unpublished manuscript, original held by Melville Jacobs Collection, University Archives, University of Washington Libraries (Microfilmed and photoduplicated copy in British Columbia Indian Language Project files, Victoria, Canada). 1952-1954

[49] Turner NJ, Bouchard R, Kennedy DID. Ethnobotany of the Okanagan-Colville Indians of British Columbia and Washington. Victoria, BC: British Columbia Provincial Museum; 1980

Chapter 10

Sustainable Pasture Management

Atanas Sevov, Chtistina Yancheva and
Yanka Kazakova

Additional information is available at the end of the chapter

http://dx.doi.org/10.5772/intechopen.72310

Abstract

Grasslands which are a major part of the global ecosystem, covering 37% of the earth's terrestrial area, have a significant contribution to food security through providing most of the energy and proteins required by the ruminants used for meat and dairy production. Grasslands are considered to have the potential to play a fundamental role in climate change mitigation, particularly regarding carbon storage and sequestration and for biodiversity preservation. This chapter provides an overview of the causes of the pasture degradation and some essential elements for sustainable management, which aims to improve the quantity and quality of pasture, mitigation of climate change and biodiversity preservation. Another point of this chapter is the grasslands with high nature value that nowadays is a top priority in the European legislation as the European Commission has confirmed that HNV farming will remain a key priority in 2014–2020. We present the situation in Bulgaria because it is one of the first member state countries that have assessed HNV regions and put funding in place to support them.

Keywords: grasslands, grass composition, perennial grasses, uncontrolled grazing, sustainable management, rotational grazing, high nature value (HNV), Bulgarian grasslands

1. Introduction

Future challenges related to the sustainable management of natural resources and investments in food production, agriculture and biotechnology research can be summarized as follows: global population growth (the population of the earth will be about 9.2 billion in 2050), global climate change and its adverse impact on agriculture [1], depletion of natural resources with significant importance for the development of world agriculture (e.g. global phosphorus deposits), food safety and security and new ethical requirements for producers.

Grasslands which are a major part of the global ecosystem, covering 37% of the earth's terrestrial area, have a significant contribution to food security through providing most of the energy and proteins required by the ruminants used for meat and dairy production. Grasslands are considered to have the potential to play a fundamental role in greenhouse gas (GHG) mitigation, particularly regarding carbon storage and sequestration. Conant et al. [2] conclude that grasslands can act as a significant carbon sink with the implementation of improved management. According to the estimation of FAO [3], global carbon stocks in grasslands is about 343 Gt C, which is about 50% more than the amount stored in world forests.

O'Mara [4] indicates that grazing management and pasture improvement have a global technical potential for mitigation of almost 1.5 Gt carbon dioxide equivalents in 2030, with additional reduction possible from the restoration of degraded lands. According to Nordborg and Röös [5], the total carbon storage potential in pastures does not exceed 0.8 tons of C per ha and year or 27 billion tons of C globally. During the last years, many researchers studied the function of grasslands as a carbon sink and the main factors affecting the storage process [2, 5–11, 46]. According to some authors [6, 9, 10, 12, 18, 22, 27], soil's grazing intensity (under- and overgrazing) can lower carbon sequestration or lead to carbon losses. These authors observed effects of grazing mediated by changes in the removal, growth, carbon allocation and flora in pastures and carbon input from ruminant excreta, which affect the amount of carbon in soils [27, 36, 40].

The results of the studies conducted by Alemu et al. [14] indicated that grazing management practices impacted greenhouse gas (GHG) intensity of beef production by affecting diet quality, animal performance and soil C change. It also emphasizes the importance of accounting for all emission sources and sinks within a beef production system when estimating its environmental impacts.

2. Effect of continuous (uncontrolled) grazing on grasslands

Formation and development of grass compositions in meadows and pastures were conditioned with the influence of soil and weather, relief, altitude, plant species interactions, microorganisms, animals and humans. All these factors are interrelated and constantly changing due to variations in the species composition and the quantitative ratio of the different species and groups. Plants in meadows and pastures are changing relatively fast under the influence of different anthropogenic pressures, which can cause both positive and negative changes.

Many high nature value pastures have been abandoned. The meadow mowing has been ceased, which leads to developing of more aggressive grass species, shrubs and trees. Wood and shrub forms begin due to uncontrollable spread or the existence of a forest near the grassland, which gradually spreads from the end to the middle of the area.

Due to the weak animal's grazing efficiency in seminatural grasslands, many of them are degraded and turned into arable land, orchards or vineyards.

This leads to the irreversible loss of diversity of plant species as well as of vertebrate and invertebrate species.

The economic status of meadows and pastures is determined by different characteristics of the grasses in them, while the most important indicator is the lawn productivity in normal climatic conditions, which depends on the soil fertility and regimen of use.

Another indicator of the economic status of meadows and pastures is the quality of green mass and hay, which is determined by the degree of acceptance by the animals, nutritional value (protein, vitamins, mineral, salts) and digestibility of the plant species which are part of the grassland.

Many plant species from different botanical families are found in natural meadows and pastures. In comparatively similar areas, soil and climatic conditions, the number of species in grasslands often exceed 50–60. Meadow and pasture grasses are divided into three groups, cereals, legumes and grasses, from other botanical families, which are referred to as "various plants."

The widest distribution among grasslands has the species from Poaceae family up to 50–90% of the grass [15, 16]. This is due to their highly competitive ability, longer shelf life and durability of unfavorable climatic and soil conditions. Cereal grasses are wanted component in the grasslands because they supply the animals with easily digestible and rich of nutritions biomass. They also protect the soil from water and wind erosion due to dense grass they develop.

Legumes have the highest nutritional value but are less widespread—very often with 6–10%. Their seldom occurrence in meadows and pastures is due to their greater rigor to the environment and the less durability of most species in the family. Only in conditions very favorable to their development, they can reach 50–60% [15, 16]. Compared to cereals and various plants (from other botanical families), legumes are less common in meadows and pastures. They are not a constant element of grasslands, and their participation is strongly influenced by climatic conditions—in wet years, the so-called clover years are more abundant, and in dry years, their involvement is insignificant. Increasing legumes is a way to improve the quality of grassy biomass, as they are rich in proteins, minerals and vitamins. Their higher number in grasslands improves the nitrogen balance of the soil and promotes the more active development of the other species.

The distribution of the various species (from other botanical families), in grasslands, is determined by the peculiarities of the environment—their participation varies from 10 to 60% [15, 16]. This group is distinguished by a great variety of species—there are about 200, with different nutritional values [15]. This group is represented in mountain meadows and high mountain pastures as well as in wet meadows and pastures.

In the grass cover of meadows and pastures, the perennial grasses of these groups prevail. One-year species rarely occur, with greater participation in degraded grasslands as well as in abandoned orchards. These species are important for early grazing in the southern and southeast parts of Bulgaria [16, 28].

Proper and regulated pasture loading is of paramount importance for ensuring quality grazing with valuable botanical composition, conservation of species diversity and longer use. Still, in many countries, free grazing is applied, which damages both the grasses density and species proportion in grasslands.

Also, soil compaction leads to a change in its physical properties, which affects the development of the grass. Species that require better aeration quickly drop out of grasslands and in their place develop more stable grasses that are less productive and inferior in quality. In this way of grazing, only 40% of the grass is used [45].

Grazing has a strong influence on grass composition. Changes in grassland under the grazing influence depend on the kind of animals, the time and the way of grazing, the soil conditions, the grassland peculiarity, etc. [44]. Grazing early in the spring can suppress some valuable species and allow domination of the weeds [16, 18].

The soil compaction increases the number of rhizomatous grasses that are less sensitive to soil aeration and reduces the participation of the demand in this regard of rhizome and high-growth bunch grasses.

In moderately wet pastures, grazing contributes to consolidation and compaction of the sward, and in damp pastures, trampling can lead to swamping and allow invasion of some weeds, casing poaching. In the dry grasslands, the steady treading leads to shattering a grass cover. The unfavorable influence of treading strongly occurs at unsystematic grazing when the animals move freely and stay for a long time in the pasture. It is stronger when the pasture was used by cattle. The grass composition in grasslands significantly changes with the grazing and species selection from the animals during grazing. Under the influence of grazing some plants, fall off the grassland. During the grazing, the animals eat almost entirely the leaves of tall grasses, which make their recovery difficult, and they are relatively quickly dropped out of the grassland. Low-growth and rosette plants are recovering faster as they retain their basal leaves and take the lead in grassland. They are well preserved in pastures and species with creeping, rooting stems and inflorescences near the soil surface (white clover, knotgrass, etc.).

In grazing, animals prefer certain plants, while others avoid. In the case of unsympathetic grazing, the species that animals avoid form seeds, and the grasses they prefer reduce their vitality and gradually fall out of grass. In rotational grazing, the influence of the animal selection during grazing is almost eliminated, while the rest of the plants were harvested, until they are re-grazed and with the practice of cutting the grass, left after the grazing, is complete in each grazing cycle. In case of free grazing, the influence of the different animals on grassland components is also more pronounced. It was known that the sheep graze the grass shallow, making it difficult to restore the common pasture grasses, but cattle partially plucked up some species during grazing.

This effect was observed in many investigations [4, 19, 20], reduced growth, tiller numbers, plant cover and changes in botanical composition.

Early spring and late autumn grazing reduce the participation of valuable pasture grasses, which have not accumulated enough reserves to overcome early grazing and survive during winter. This grazing leads to an increase in the participation of the first developed annual species that grow up by seeds.

In cattle grazing, hard stools have some adverse effects. The plants below them suffocate and die, and some nitrophilic species develop around them. The larvae of some insects and helminths that cause animal diseases develop in the field. The urine, rich in nitrogen, favors the development of valuable species.

Intensive grazing is depleting the soil, despite the fact that part of the grass-fed nutrients was restored to the soil with animal stools. Soil degradation reduces the participation of valuable pasture grasses which are demanding for the presence of nutrients and increases the involvement of the low-productive, medium-quality, densely tufted grasses.

The negative consequences of nonsystemic, uncontrolled grazing can only overcome by introducing an appropriate grazing regime. Systematic and organized grazing (regular or parcel) would help to preserve species diversity and grass density.

3. Grazing management

Grazing management—combining animal, plant, soil and other environmental components and the grazing methods by which the system is managed to achieve specific goals—improved pasture condition, higher forage yields and animal production with ecological concern [16].

The sustainable grazing management includes a proper stocking rate, livestock type and recovery time for grass regrowth after grazing. It is important to consider the effect of grazing management on pasture growth, tiller density, pasture quantity and quality and soil properties. Many factors affect quantity and quality of pastures like farm topography, weather variation among the seasons, botanical composition, herbage cover, stocking rate, seasonal grazing, anti-quality compounds in grasses and application of different practices [23, 24].

Rotational grazing was a component of the institutional and scientific response to severe rangeland degradation at the turn of the twentieth century, and it has since become the professional norm for grazing management [25, 31].

What is rotational grazing—all cases in which only one part of pasture is grazed while all other parts rest? That means that the pasture is divided into a certain number of small areas (paddocks), and the livestock can use only one of them. In this case, the grazing animals are moved from one paddock to another and are thus forced to graze much of the grass. In all other (resting) parts, the grass can renew its energy reserves deepen the root system and in the future time to give a maximum production.

Why use rotational grazing? All over the world, people with livestock and grazing land can benefit from rotational grazing. This has some advantages, called benefits such as economic benefits, time savings, environmental benefits, esthetics and human health benefits, better animal health, etc.

A rotational grazing system is preferred in pasture-based animal production because meat from cows and lambs has better quality with less fat, more vitamin E [26, 29] and higher levels of omega-3 and conjugated linoleic fatty acids than grain-finished products [17, 21, 30].

The fundamental advantage for the animals in grazing systems is that the livestock in pastures is healthier than these housed in confinement. The key of sufficient rotational grazing is determination of the number and size of paddocks, water supply for livestock and fence type. Determination of the suitable number of paddocks depends on the time required for grass regrowth (**Table 1**) and grazing period that is varied from 4 to 6 days.

Many authors in their publications present different grazing management models with the consideration of periods of strong growth and animal pressure [7, 43]. Many farmers in countries with hill pastures applied adopted regime with 3 days per paddock and high stocking rate [23].

Jacobo [32] observed that productivity and sustainability might be compatible by replacing continuous with rotational grazing. The reason is that rotational grazing promoted functional groups composed of high forage value species and reduced bare soil through the accumulation of plant residues. These changes indicate an improvement in rangeland condition and in carrying capacity.

These results are relevant to the other authors. As an example, Pavlů et al. [13] studied continuous stocking and rotational grazing. On the base of the databases, authors conclude that vegetation varied as a result of time and differences between treatments. Several prostrate dicotyledonous species (*Trifolium repens* L., *Taraxacum* sp., *Bellis perennis* and *Leontodon autumnalis*) increased under continuous stocking. This treatment also promoted the growth of the perennial grass *Lolium perenne* L., which was able to cope with frequent defoliation. Tall grasses sensitive to frequent defoliation (*Poa trivialis* L., *Holcus mollis* L., *Alopecurus pratensis* L., *Dactylis glomerata* L. and *Elytrigia repens* L.) were more abundant in rotationally grazed paddocks. Species diversity was not significantly influenced by the different grazing systems. The decrease in the potential sward height under continuous stocking revealed the replacing of tall dominants by lower species. Information about pasture management should, therefore, involve not only grazing intensity but also the grazing system used.

The new opportunity to improve the management and welfare of extensively produced beef cattle is to combine technologies for monitoring the spatial behavior of livestock with technologies that monitor pasture availability. According to Manning et al. [33], the Global Navigation Satellite System (GNSS) technology could determine livestock grazing preference and hence improve management and paddock utilization. The cattle behavior changed, highlighting how technologies that monitor these two variables may be used in the future as management tools to assist producers better manage cattle and to manipulate grazing intensity and paddock utilization.

Species	**Cool weather**	**Hot weather**
Cool season grasses	14	35–50
Warm season grasses	35–40	21
Legumes	21–28	21–28

Source: Blanchet et al. [31].

Table 1. Optimal rest period for forage species in days.

Sustainable pasture management including the application of new technologies has several environmental advantages over tilled land—significantly decrease soil erosion, require minimal pesticides and fertilizer usage and reduce the amount of barnyard runoff. This leads to the conclusion that, taking advantage of wildlife, we can also increase the pasture productivity.

4. High nature value (HNV) grasslands in Bulgaria

By definition high nature value (HNV) farmland represents areas where "agriculture is a major (usually the dominant) land use and where that agriculture supports, or is associated with, either a high species and habitat diversity or the presence of species of European, and/or national, and/or regional conservation concern, or both" [34, 35]. The majority of HNV farmland and in Europe comprises seminatural pastures, meadows and orchards as well as various landscape elements [35, 38]. Around one-third of the agricultural area in Bulgaria is potentially of high nature value, and the most significant share of it is seminatural pastures and meadows [41, 45]. The figures below visualize high nature value grasslands in Bulgaria: flower-rich meadows in Elena municipality (**Figure 1**), species-rich pastures in Central Balkans (**Figure 2**) and species-rich pastures in Eastern Stara Planina (**Figure 3**).

HNV grasslands are of particular importance for nature conservation and the European ecological network of protected areas of Natura 2000. There are 18 habitats of natural and seminatural grassland ecosystems in the Bulgarian Natura 2000 sites, which cover between 15 and 20% of their territory [37].

Key features of the HNV farming systems are the low inputs, low outputs and high labor requirements usually resulting in a significant number of species and structural diversity in space and time [38]. The practices most often associated with HNV pastures and meadows are extensive grazing as presented on **Figure 4** and cutting hay (mowing) once or twice per year. **Figure 5** shows traditional hay storage still preserved in Western Stara Planina.

Figure 1. Flower-rich meadows in Elena municipality (June 2012, Y. Kazakova).

Figure 2. Species-rich pastures in Central Balkans (July 2016, Y. Kazakova).

Figure 3. Species-rich pastures in Eastern Stara Planina (June 2012, Y. Kazakova).

However, the modernization of agriculture inevitably leads to the intensification of the traditional practices and decrease in the high nature value. For example, over 90% of the grassland habitats in the European ecological network Natura 2000 are in unfavorable conservation

Figure 4. Extensive sheep grazing in Central Stara Planina (July 2016, Y. Kazakova).

Figure 5. Traditional hay storage still preserved in Western Stara Planina (November 2014, Y. Kazakova).

status [37]. The two extreme examples are the loss of HNV grasslands due to conversion to intensive meadows or even arable land and the abandonment of farming in areas unsuitable for intensification. **Figure 6** presents scrub overgrowth and closure of landscapes in abandoned grasslands in Western Stara Planina.

The trend is best revealed by the statistical data on grasslands in Bulgaria, presented in **Figure 7** (BANCIK MAF, 2000–2016). The total area of grasslands in Natura 2000 was just over 1.8 million hectares. In 2015, it was down to 1.36 million hectares, a decrease of 24% [41].

Overall, grasslands cover around one-third of the agricultural area in Bulgaria. In the agricultural land use surveys, they are divided into four grassland groups:

1. Permanent productive meadows, which can be natural or planted for longer than 6 years and can be used either for mowing or for grazing. Their area decreased by 38% from 2000 to 2015. Due to their high productivity, they are often converted to arable land.

Figure 6. Grasslands abandonment leads to scrub overgrowth and closure of landscapes in Western Stara Planina (April 2015, Y. Kazakova).

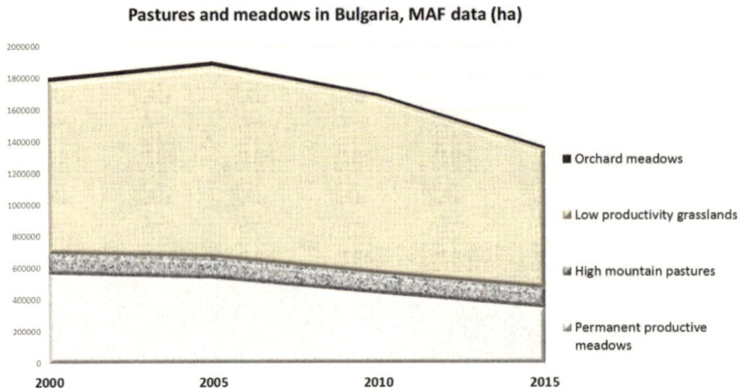

Figure 7. Total area of grasslands in Bulgaria (2000–2015).

2. High mountain pastures are located at altitudes between 1000 and 1500 m a.s.l. and are used for summer grazing of livestock. Their area is most stable in comparison to other grassland groups—it decreased by only 6% from 2000 to 2015.
3. Low productivity grasslands—usually used for grazing.
4. Due to their low productivity, they are never mown. They decreased by 19% from 2000 to 2015 mostly due to an abandonment of farming (**Figure 7**).
5. Orchard meadows, which are permanent productive pastures in orchards with less than 100 trees per hectare. Their area decreased the most, by 46%, from 2000 to 2015.

Another negative tendency for the loss of HNV grasslands is their sale for development. The extensive land use and the species-rich grasslands, as well as the site's characteristics, often create landscapes that are attractive for tourists as shown in **Figure 8**. This creates development pressure, and the values that attracted visitors ultimately were lost.

Figure 8. Grasslands for sale: the attractive high nature value landscapes stimulate tourism development (June 2012, Y. Kazakova).

When HNV farmlands were first identified in 2007 in Bulgaria, the area of HNV grasslands was estimated at 951,256 ha [39]. Only 5 years later, in 2012, the HNV grassland area decreased to 809,530 ha [37]. Even if there were some methodological differences, the decreasing trend is unquestionable.

To preserve and maintain grassland areas of high nature value and the associated species, measures financed under the Bulgarian Rural Development Programme (2014–2020) [42] are being undertaken to promote or restore traditional management practices for seminatural grassland, as follows:

- Keeping the density of livestock units at 0.3–1 LU/ha according to the natural, climatic and soil conditions to ensure the good ecological status of meadows and pastures and maintenance of a permanent grass cover.
- A ban on the use of mineral fertilizers and pesticides.
- Cleaning of undesirable grass and shrub vegetation.
- Consecutive grazing.

Overall, HNV grasslands in Bulgaria require targeted policy support and improved management both from agricultural and conservation point of view to improve the current situation where the forage resources are decreasing because of the loss of grassland area; the natural quality of the remaining grasslands is also declining due to the intensification or abandonment of extensive, low-input practices.

Author details

Atanas Sevov[1]*, Chtistina Yancheva[1] and Yanka Kazakova[2]

*Address all correspondence to: asevov@yahoo.com

1 Agricultural University, Plovdiv, Bulgaria

2 University of National and World Economy, Sofia, Bulgaria

References

[1] Scollan ND, Enser M, Richardson RI, Wood JD. Effect of forage legumes on the fatty acid composition of beef. The Proceedings of the Nutrition Society. 2002;**61**(3A):99A

[2] Conant RT, Paustian K, Elliott ET. Grassland management and conversion into grassland: Effects on soil carbon. Ecological Applications. 2001;**11**(2):343-355

[3] FAO. Challenges and opportunities for carbon sequestration in grassland systems—A technical report on grassland management and climate change mitigation. Integrated Crop Management. 2010;**9**

[4] O'Mara FP. The role of grasslands in food security and climate change. Annals of Botany. 2012;**110**(6):1263-1270. DOI: 10.1093/aob/mcs209

[5] Nordborg M, Röös E. Holistic management—A critical review of Allan Savory's grazing method. SLU, Swedish University of Agricultural Sciences & Chalmers. 2016; ISBN: 978-91-576-9424-9

[6] Deng L et al. Effects of grazing exclusion on carbon sequestration in China's grassland. Earth-Science Reviews. 2017;**173**:84-95. DOI: 10.1016/j.earscirev.2017.08.008

[7] Sheath GW, Carlson WT. Impact of cattle treading on hill land 1. Soil damage patterns and pasture status. New Zealand Journal of Agricultural Research. 1998;**41**:271-278

[8] Schonbach P, Wan HW, Gierus M, Bai YF, Muller K, Lin LJ, Susenbeth A, Taube F. Grassland responses to grazing: Effects of grazing intensity and management system in an inner Mongolian steppe ecosystem. Plant and Soil. 2011;**340**(1-2):103-115

[9] Wang S, Wilkes A, Zhang Z, Chang X, Lang R, Wang Y, Niu H. Management and land use change effects on soil carbon in northern China's grasslands: A synthesis. Agriculture, Ecosystems & Environment. 2011;**142**(3-4):329-340. DOI: 10.1016/j.agee.2011.06.002

[10] Wang D, Wu GL, Zhu YJ, Shi ZH. Grazing exclusion effects on above- and below-ground C and N pools of typical grassland on the Loess Plateau (China). Catena. 2014;**123**:113-120. DOI: 10.1016/j.catena.2014.07.018

[11] Zavaleta ES, Shaw MR, Chiariello NR, Mooney HA, Field CB. Additive effects of simulated climate changes, elevated CO_2, and nitrogen deposition on grassland diversity. Proceedings of the National Academy of Sciences. 2003;**100**(13):7650-7654

[12] Liebig MA, Morgan JA, Reeder JD, Ellert BH, Gollany HT, Schuman GE. Greenhouse gas contributions and mitigation potential of agricultural practices in northwestern USA and western Canada. Soil & Tillage Research. 2005;**83**:25-52. DOI: 10.1016/j.still.2005.02.008

[13] Pavlů V, Hejcman M, Pavlů L, Gaisler J. Effect of rotational and continuous grazing on vegetation of an upland grassland in the Jizerské Hory Mts., Czech Republic. Folia Geobotanica. 2003;**38**(1):21-34

[14] Alemu AW et al. Assessment of grazing management on farm greenhouse gas intensity of beef production systems in the *Canadian prairies* using life cycle assessment. Agricultural Systems. 2017;**158**:1-13. DOI: 10.1016/j.agsy.2017.08.003

[15] Ministry of Environment, Ministry of Agriculture. Sustainable land management. Ministry of Environment—Bulgarian Government Publisher; 2007

[16] Yancheva C, Angelova S, Koeva R. Management of steppe habitats in pastures of the region of Kaliakra reserve in Bulgaria. In: XIX-th General Meeting of EGF in Multifunction Grasslands; La Roshelle, France. 2002. pp. 860-862

[17] Yang A, Lanari MC, Brewster M, Tume RK. Lipid stability and meat colour of beef from pasture- and grain-fed cattle with or without vitamin E supplement. Meat Science. 2002; **60**(1):41-50. DOI: 10.1016/S0309-1740(01)00103-6

[18] Wang KB, Deng L, Ren ZP, Li JP, Shangguan ZP. Grazing exclusion significantly improves grassland ecosystem C and N pools in a desert steppe of China. Catena. 2016;**137**:441-448. DOI: 10.1016/j.catena.2015.10.018

[19] Garden DL, Lodge GM, Friend DA, Dowling PM, Orchard BA. Effects of grazing management on botanical composition of native grass-based pastures in temperate south-east Australia. Australian Journal of Experimental Agriculture. 2000;**40**:225-245. DOI: 10.1071/EA98010

[20] Garden DL, Dowling PM, Eddy DA, Nicol HI. The influence of climate, soil, and management on the composition of native grass pastures on the central, southern, and Monaro tablelands of New South Wales. Australian Journal of Agricultural Research. 2001;**52**(9):925-936. DOI: 10.1071/AR98184

[21] Ripoll G, Albertí P, Casasús I, Blanco M. Instrumental meat quality of veal calves reared under three management systems and color evolution of meat stored in three packaging systems. Meat Science. 2013;**93**:336-343. DOI: 10.1016/j.meatsci.2012.09.012

[22] Shaw MR, Zavaleta ES, Chiariello NR, Cleland EE, Mooney HA, Field CB. Grassland responses to global environmental changes suppressed by elevated CO_2. Science. 2002; **298**(5600):1987-1990

[23] King WM, Rennie GM, Devantier B, Hoogendoorn CJ. Impacts of grazing management on hill country pastures: Principles and practices. Hill Country – Grassland Research and Practice. 2016;**16**:203-212

[24] Lambert MG, Clark DA, Litherland AJ. Advances in pasture management for animal productivity and health. New Zealand Veterinary Journal. 2004;**52**:311-319. DOI: 10.1080/00480169.2004.36447

[25] Briske DD, Derner JD, Brown JR, Fuhlendorf SD, Teague WR, Havstad KM, Willms WD. Rotational grazing on rangelands: Reconciliation of perception and experimental evidence. Rangeland Ecology & Management. 2008;**61**(1):3-17. DOI: 10.2111/06-159R.1

[26] Descalzo AM, Insani EM, Biolatto A, Sancho AM, Garcia PT, Pensel NA. Influence of pasture or grain-based diets supplemented with vitamin E on antioxidant/oxidative balance of argentine beef. Meat Science. 2005;**70**:35-44. DOI: 10.1016/j.meatsci.2004.11.018

[27] Rice CW, Owensby CE. The effects of fire and grazing on soil carbon in rangleands. The potential of US grazing lands to sequester carbon and mitigate the greenhouse effect. 2001. DOI: 10.1201/9781420032468.ch13

[28] Yancheva C, Angelova S, Guteva Ya, Mihovski Tz. Steppe habitats in the pastures of the region of Kaliakra reserve and their preservation. Journal of Mountain Agriculture on the Balkans. 2007:104-109

[29] Fraser TJ, Rowarth JS. Legumes, herbs or grass for lamb performance. Proceedings—NZ Grassland Association. 1996;**58**:49-52

[30] Gutierrez-Pena R et al. Carcass composition and meat quality of pasture-raised Mallorquina sheep in Balearic Islands. Grassland resources for extensive farming systems in marginal lands: Major drivers and future scenarios. In: Proceedings of the 19th Symposium of the European Grassland Federation; 7-10 May 2017; Alghero, Italy. CNR-ISPAAM; 2017

[31] Blanchet K, Moechnig H, DeJong-Hughes J. Grazing systems planning guide. University of Minnesota Extension Service; 2000

[32] Jacobo EJ, Rodríguez AM, Bartoloni N, Deregibus VA. Rotational grazing effects on rangeland vegetation at a farm scale. Rangeland Ecology & Management. 2006;**59**(3):249-257. DOI: 10.2111/05-129R1.1

[33] Manning J et al. The behavioural responses of beef cattle (*Bos taurus*) to declining pasture availability and the use of GNSS technology to determine grazing preference. Agriculture. 2017;**7**(5):45. DOI: 10.3390/agriculture7050045

[34] Andersen E, Baldock D, Bennet H, Beaufoy G, Bignal E, Brower F, Elbersen B, Eiden G, Godeschalk F, Jones G, McCracken DI, Nieuwenhuizen W, van Eupen M, Hennekes S, Zervas G. Developing a High Nature Value Farming Area Indicator. Consultancy Report to the EEA. Copenhagen: European Environment Agency; 2003

[35] Beaufoy G, Jones G, Oppermann R. High Nature Value Farming in Europe. 35 European Countries – Experiences and Perspectives. Ubstadt-Weiher: Verlag Regionalkultur; 2012

[36] Soussana JF, Tallec T, Blanfort V. Mitigating the greenhouse gas balance of ruminant production systems through carbon sequestration in grasslands. Animal. 2010;**4**:334-350. DOI: 10.1017/S1751731109990784

[37] Ministry of Environment. National Priority Action Framework for Natura 2000 in Bulgaria for the 2014-2020 period. Ministry of Agriculture—Bulgarian Government Publisher; 2014

[38] Oppermann R, Paracchini ML. HNV Farming—Central to European Cultural Landscapes and Biodiversity. 35 European Countries—Experiences and Perspectives. Ubstadt-Weiher: Verlag Regionalkultur; 2012

[39] Ministry of Agriculture. Bulgarian Rural Development Programme for the 2007-2013 Programming Period. Ministry of Agriculture—Bulgarian Government Publisher; 2007

[40] Liu S, Schleuss PM, Kuzyakov Y. Carbon and nitrogen losses from soil depend on degradation of Tibetan Kobresia pastures. Land Degradation and Development. 2016;**28**(4):1253-1262. DOI: 10.1002/ldr.2522

[41] Ministry of Agriculture. 2000-2016 Agristatistics Annual BANCIK Reports on the Land Cover and Land Use in Bulgaria. Ministry of Agriculture—Bulgarian Government Publisher; 2016

[42] Ministry of Agriculture. Bulgarian Rural Development Programme for the 2014-2020 programming period. Ministry of Agriculture—Bulgarian Government Publisher; 2014

[43] Pande TN, Valentine I, Betteridge K, MacKay A, Horne D. Pasture damage and regrowth from cattle treading. Proceedings of the New Zealand Grassland Association. 2000;**62**: 155-160

[44] Shakhane LM et al. Changes in botanical composition on three farmlets subjected to different pasture and grazing management strategies. Animal Production Science. 2013; **53**(8):670-684

[45] Stefanova V, Kazakova Y. HNV Farming in Bulgaria. High Nature Value Farming in Europe. 35 European Countries—Experiences and Perspectives. Ministry of Agriculture - Bulgarian government publisher; 2012

[46] Zhou ZY, Li FR, Chen SK, Zhang HR, Li GD. Dynamics of vegetation and soil carbon and nitrogen accumulation over 26 years under controlled grazing in a desert shrubland. Plant and Soil. 2011;**341**(1-2):257-268